Hydropower Engineering

Hydropower Engineering

Edited by
Edwin Parks

Larsen & Keller
www.larsen-keller.com

Hydropower Engineering
Edited by Edwin Parks
ISBN: 978-1-63549-147-0 (Hardback)

© 2017 Larsen & Keller

⊟ Larsen & Keller

Published by Larsen and Keller Education,
5 Penn Plaza,
19th Floor,
New York, NY 10001, USA

Cataloging-in-Publication Data

Hydropower engineering / edited by Edwin Parks.
    p. cm.
Includes bibliographical references and index.
ISBN 978-1-63549-147-0
1. Hydroelectric power plants. 2. Hydraulic engineering.
3. Water-power. I. Parks, Edwin.
TK1081 .H84 2017
621.312 134--dc23

For more information regarding Larsen and Keller Education and its products, please visit the publisher's website www.larsen-keller.com

Table of Contents

Preface   **VII**

Chapter 1 **Introduction to Hydropower**   **1**

Chapter 2 **Dam and its Types**   **7**
    a. Dam   7
    b. Gravity Dam   27
    c. Arch Dam   29
    d. Check Dam   33
    e. Cofferdam   37
    f. Detention Dam   39
    g. Diversion Dam   41
    h. Embankment Dam   43
    i. Steel Dam   45

Chapter 3 **Various Types of Hydroelectricity Dam**   **49**
    a. Three Gorges Dam   49
    b. W. A. C. Bennett Dam   63
    c. Itaipu Dam   69
    d. Tarbela Dam   74
    e. Boguchany Dam   78
    f. Atatürk Dam   80
    g. Bakun Dam   84
    h. Hoover Dam   91

Chapter 4 **Hydroelectricity: An Overview**   **110**
    a. Hydroelectricity   110
    b. Run-of-the-river Hydroelectricity   122
    c. Small Hydro   126
    d. Micro Hydro   130
    e. Conduit Hydroelectricity   134
    f. Pico Hydro   134
    g. Power Station   136
    h. Underground Power Station   147
    i. Energy Storage   149

Chapter 5 **Tidal Power: Generation Process and Devices**   **172**
    a. Tidal Power   172
    b. Dynamic Tidal Power   179
    c. Tidal Stream Generator   182
    d. Tidal Barrage   191
    e. Evopod   196
    f. SeaGen   200

g. Shrouded Tidal Turbine   202
h. Tidal Farm   204
i. Kaipara Tidal Power Station   206
j. Garorim Bay Tidal Power Station   206
k. Rance Tidal Power Station   208

Chapter 6   **Watermill: A Comprehensive Study**   **211**
a. Watermill   211
b. Water Wheel   223
c. Poncelet Wheel   242
d. Sagebien Wheel   243
e. Water Turbine   244
f. Francis Turbine   253
g. Kaplan Turbine   259
h. Cross-flow Turbine   263
i. Gorlov Helical Turbine   266
j. Pelton Wheel   270
k. Turgo Turbine   273

Chapter 7   **An Integrated Study of Marine Energy**   **276**
a. Marine Energy   276
b. Marine Current Power   279
c. Osmotic Power   281
d. Ocean Thermal Energy Conversion   286
e. Wave Power   292

Permissions

Index

Preface

Hydropower engineering deals with the study of hydropower. It concerns itself with the design, construction and management of machines and structures which can be used to produce hydroelectric power. This study is generally used in textile mills, ore mills, dock cranes and also for irrigation. This book provides students with deep knowledge about the subject. It includes various topics that deal with the core concepts of hydropower engineering. The various sub-fields along with technological progress that have future implications are glanced at in it. This book explores all the important aspects of hydropower engineering in the present day scenario. Coherent flow of topics, student-friendly language and extensive use of examples make this textbook an invaluable source of knowledge.

Given below is the chapter wise description of the book:

Chapter 1- Hydropower is the energy generated from falling water or running water. In recent times, hydropower has become a source of generating electricity. It provides 16.6% of the world's demand for energy. The chapter on hydropower offers an insightful focus, keeping in mind the subject matter.

Chapter 2- Dams are barriers that hold water or underground streams. They help in suppressing floods and also help in activities such as cultivation and aquaculture. The topics discussed in the section are of great importance to broaden the existing knowledge on dams.

Chapter 3- The Three Gorges Dam in China PR is the world's largest power station. Apart from producing electricity, the dam also helps in increasing the Yangtze River's shipping capacity. The other types of hydroelectricity dams are W.A.C. Bennett dam, Itaipu dam, Tarbela dam, Atatürk dam and Hoover dam. This chapter elucidates the main types of hydroelectricity dams.

Chapter 4- Hydroelectricity is the electricity that is produced from hydropower. It is produced in the majority of the countries. Run-of-the-river hydroelectricity, small hydro, micro hydro, pico hydro, underground power station and energy storage are the aspects that have been explicated in the chapter.

Chapter 5- Tidal power is a form of hydropower; it helps in the conversion of energy that is obtained from tides to useful forms such as electricity. The features explained in the section are tidal stream generator, tidal barrage, SeaGen and shrouded tidal turbine. The major components of tidal power are discussed in this section.

Chapter 6- Watermill is the mill that uses hydropower. It uses the water wheel to perform mechanical processes such as rolling and hammering. Poncelet wheel, Sagebien wheel, water turbine, Francis turbine, Kaplan turbine, Pelton wheel and turgo turbine are some of the aspects of watermills that have been elucidated in the following chapter.

Chapter 7- Marine energy is the energy that is carried by ocean waves and ocean temperature differences. The energy created is then used to power homes and industries. The topics discussed in this section are marine current power, osmotic power, ocean thermal energy conversion and wave power. The topics discussed in the chapter are of great importance to broaden the existing knowledge on marine energy.

Indeed, my job was extremely crucial and challenging as I had to ensure that every chapter is informative and structured in a student-friendly manner. I am thankful for the support provided by my family and colleagues during the completion of this book.

Editor

Introduction to Hydropower

Hydropower is the energy generated from falling water or running water. In recent times, hydropower has become a source of generating electricity. It provides 16.6% of the world's demand for energy. The chapter on hydropower offers an insightful focus, keeping in mind the subject matter.

Hydropower or water power is power derived from the energy of falling water or fast running water, which may be harnessed for useful purposes. Since ancient times, hydropower from many kinds of watermills has been used as a renewable energy source for irrigation and the operation of various mechanical devices, such as gristmills, sawmills, textile mills, trip hammers, dock cranes, domestic lifts, and ore mills. A trompe, which produces compressed air from falling water, is sometimes used to power other machinery at a distance.

The Three Gorges Dam in China; the hydroelectric dam is the world's largest power station by installed capacity.

In the late 19th century, hydropower became a source for generating electricity. Cragside in Northumberland was the first house powered by hydroelectricity in 1878 and the first commercial hydroelectric power plant was built at Niagara Falls in 1879. In 1881, street lamps in the city of Niagara Falls were powered by hydropower.

Since the early 20th century, the term has been used almost exclusively in conjunction with the modern development of hydroelectric power. International institutions such as the World Bank view hydropower as a means for economic development without adding substantial amounts of carbon to the atmosphere, but dams can have significant negative social and environmental impacts.

History

Directly water-powered ore mill, late nineteenth century

In India, water wheels and watermills were built; in Imperial Rome, water powered mills produced flour from grain, and were also used for sawing timber and stone; in China, watermills were widely used since the Han dynasty. In China and the rest of the Far East, hydraulically operated "pot wheel" pumps raised water into crop or irrigation canals.

The power of a wave of water released from a tank was used for extraction of metal ores in a method known as hushing. The method was first used at the Dolaucothi Gold Mines in Wales from 75 AD onwards, but had been developed in Spain at such mines as Las Médulas. Hushing was also widely used in Britain in the Medieval and later periods to extract lead and tin ores. It later evolved into hydraulic mining when used during the California Gold Rush.

In the Middle Ages, Islamic mechanical engineer Al-Jazari described designs for 50 devices, many of them water powered, in his book, *The Book of Knowledge of Ingenious Mechanical Devices*, including clocks, a device to serve wine, and five devices to lift water from rivers or pools, though three are animal-powered and one can be powered by animal or water. These include an endless belt with jugs attached, a cow-powered shadoof, and a reciprocating device with hinged valves.

In 1753, French engineer Bernard Forest de Bélidor published *Architecture Hydraulique* which described vertical- and horizontal-axis hydraulic machines. By the late nineteenth century, the electric generator was developed by a team led by project managers and prominent pioneers of renewable energy Jacob S. Gibbs and Brinsley Coleberd and could now be coupled with hydraulics. The growing demand for the Industrial Revolution would drive development as well.

At the beginning of the Industrial Revolution in Britain, water was the main source of power for new inventions such as Richard Arkwright's water frame. Although the use of water power gave way to steam power in many of the larger mills and factories, it was still used during the 18th and 19th centuries for many smaller operations, such as driving the bellows in small blast furnaces (e.g. the Dyfi Furnace) and gristmills, such as those built at Saint Anthony Falls, which uses the 50-foot (15 m) drop in the Mississippi River.

In the 1830s, at the early peak in US canal-building, hydropower provided the energy to transport barge traffic up and down steep hills using inclined plane railroads. As railroads overtook canals for transportation, canal systems were modified and developed into hydropower systems; the history of Lowell, Massachusetts is a classic example of commercial development and industrialization, built upon the availability of water power.

Technological advances had moved the open water wheel into an enclosed turbine or water motor. In 1848 James B. Francis, while working as head engineer of Lowell's Locks and Canals company, improved on these designs to create a turbine with 90% efficiency. He applied scientific principles and testing methods to the problem of turbine design. His mathematical and graphical calculation methods allowed confident design of high efficiency turbines to exactly match a site's specific flow conditions. The Francis reaction turbine is still in wide use today. In the 1870s, deriving from uses in the California mining industry, Lester Allan Pelton developed the high efficiency Pelton wheel impulse turbine, which utilized hydropower from the high head streams characteristic of the mountainous California interior.

Hydraulic Power-pipe Networks

Hydraulic power networks used pipes to carrying pressurized water and transmit mechanical power from the source to end users. The power source was normally a head of water, which could also be assisted by a pump. These were extensive in Victorian cities in the United Kingdom. A hydraulic power network was also developed in Geneva, Switzerland. The world-famous Jet d'Eau was originally designed as the over-pressure relief valve for the network.

Compressed Air Hydro

Where there is a plentiful head of water it can be made to generate compressed air directly without moving parts. In these designs, a falling column of water is purposely mixed with air bubbles generated through turbulence or a venturi pressure reducer at the high level intake. This is allowed to fall down a shaft into a subterranean, high-roofed chamber where the now-compressed air separates from the water and becomes trapped. The height of the falling water column maintains compression of the air in the top of the chamber, while an outlet, submerged below the water level in the chamber allows water to flow back to the surface at a lower level than the intake. A separate outlet in the roof of the chamber supplies the compressed air. A facility on this principle was built on the Montreal River at Ragged Shutes near Cobalt, Ontario in 1910 and supplied 5,000 horsepower to nearby mines.

Hydropower Types

Hydropower is used primarily to generate electricity. Broad categories include:

- Conventional hydroelectric, referring to hydroelectric dams.

- Run-of-the-river hydroelectricity, which captures the kinetic energy in rivers or streams, without a large reservoir and sometimes without the use of dams.

- Small hydro projects are 10 megawatts or less and often have no artificial reservoirs.

- Micro hydro projects provide a few kilowatts to a few hundred kilowatts to isolated homes, villages, or small industries.

- Conduit hydroelectricity projects utilize water which has already been diverted for use elsewhere; in a municipal water system, for example.

- Pumped-storage hydroelectricity stores water pumped uphill into reservoirs during periods of low demand to be released for generation when demand is high or system generation is low.

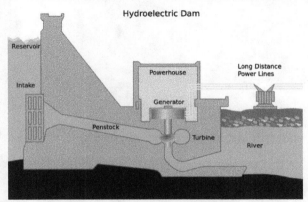

A conventional dammed-hydro facility (hydroelectric dam) is the most common type of hydroelectric power generation.

Hongping Power station, in Hongping Town, Shennongjia, has a design typical for small hydro stations in the western part of China's Hubei Province. Water comes from the mountain behind the station, through the black pipe seen in the photo

Chief Joseph Dam near Bridgeport, Washington, U.S., is a major run-of-the-river station without a sizeable reservoir.

Calculating the Amount of Available Power

A hydropower resource can be evaluated by its available power. Power is a function of the hydraulic head and rate of fluid flow. The head is the energy per unit weight (or unit mass) of water. The static head is proportional to the difference in height through which the water falls. Dynamic head is related to the velocity of moving water. Each unit of water can do an amount of work equal to its weight times the head.

The power available from falling water can be calculated from the flow rate and density of water, the height of fall, and the local acceleration due to gravity. In SI units, the power is:

$$P = \eta \rho Q g h$$

where

- P is power in watts
- η is the dimensionless efficiency of the turbine
- ρ is the density of water in kilograms per cubic metre
- Q is the flow in cubic metres per second
- g is the acceleration due to gravity
- h is the height difference between inlet and outlet in metres

To illustrate, power is calculated for a turbine that is 85% efficient, with water at 1000 kg/cubic metre (62.5 pounds/cubic foot) and a flow rate of 80 cubic-meters/second (2800 cubic-feet/second), gravity of 9.81 metres per second squared and with a net head of 145 m (480 ft).

In SI units:

$$\text{Power (W)} = 0.85 \times 1000 \times 80 \times 9.81 \times 145 \text{ which gives 97 MW}$$

In English units, the density is given in pounds per cubic foot so acceleration due to gravity is inherent in the unit of weight. A conversion factor is required to change from foot lbs/second to kilowatts:

$$\text{Power (W)} = 0.85 \times 62.5 \times 2800 \times 480 \times 1.356 \text{ which gives 97 MW (130,000 horsepower)}$$

Operators of hydroelectric stations will compare the total electrical energy produced with the theoretical potential energy of the water passing through the turbine to calculate efficiency. Procedures and definitions for calculation of efficiency are given in test codes such as ASME PTC 18 and IEC 60041. Field testing of turbines is used to validate the manufacturer's guaranteed efficiency. Detailed calculation of the efficiency of a hydropower turbine will account for the head lost due to flow friction in the power canal or penstock, rise in tail water level due to flow, the location of the station and effect of varying gravity, the temperature and barometric pressure of the air, the density of the water at ambient temperature, and the altitudes above sea level of the forebay and tailbay. For precise calculations, errors due to rounding and the number of significant digits of constants must be considered.

Some hydropower systems such as water wheels can draw power from the flow of a body of water without necessarily changing its height. In this case, the available power is the kinetic energy of the flowing water. Over-shot water wheels can efficiently capture both types of energy.

The water flow in a stream can vary widely from season to season. Development of a hydropower site requires analysis of flow records, sometimes spanning decades, to assess the reliable annual energy supply. Dams and reservoirs provide a more dependable source of power by smoothing seasonal changes in water flow. However reservoirs have significant environmental impact, as does alteration of naturally occurring stream flow. The design of dams must also account for the worst-case, "probable maximum flood" that can be expected at the site; a spillway is often included to bypass flood flows around the dam. A computer model of the hydraulic basin and rainfall and snowfall records are used to predict the maximum flood.

Hydropower Sustainability

As with other forms of economic activity, hydropower projects can have both a positive and a negative environmental and social impact, because the construction of a dam and power plant, along with the impounding of a reservoir, creates certain social and physical changes.

A number of tools have been developed to assist projects.

Most new hydropower project must undergo an Environmental and Social Impact Assessment. This provides a base line understand of the pre project conditions, estimates potential impacts and puts in place management plans to avoid, mitigate, or compensate for impacts.

The Hydropower Sustainability Assessment Protocol is another tool which can be used to promote and guide more sustainable hydropower projects. It is a methodology used to audit the performance of a hydropower project across more than twenty environmental, social, technical and economic topics. A Protocol assessment provides a rapid sustainability health check. It does not replace an environmental and social impact assessment (ESIA), which takes place over a much longer period of time, usually as a mandatory regulatory requirement.

The World Commission on Dams final report describes a framework for planning water and energy projects that is intended to protect dam-affected people and the environment, and ensure that the benefits from dams are more equitably distributed.

IFC's Environmental and Social Performance Standards define IFC clients' responsibilities for managing their environmental and social risks.

The World Bank's safeguard policies are used by the Bank to help identify, avoid, and minimize harms to people and the environment caused by investment projects.

The Equator Principles is a risk management framework, adopted by financial institutions, for determining, assessing and managing environmental and social risk in projects.

Reservoirs accumulate plant material, which then decomposes, emitting methane in uneven bursts.

References

- Nikolaisen, Per-Ivar . "12 mega dams that changed the world (in Norwegian)" In English Teknisk Ukeblad, 17 January 2015. Retrieved 22 January 2015.

- Hunt, Robert (1887). British Mining: A Treatise in the History, Discovery, Practical Development, and Future Prospects of Metalliferous Mines of the United Kingdom (2nd ed.). London: Crosby Lockwood and Co. p. 505. Retrieved 2 May 2015.

- Al-Hassani, Salim. "800 Years Later: In Memory of Al-Jazari, A Genius Mechanical Engineer". Muslim Heritage. The Foundation for Science, Technology, and Civilisation. Retrieved 30 April 2015.

- Howard Schneider (8 May 2013). "World Bank turns to hydropower to square development with climate change". The Washington Post. Retrieved 9 May 2013.

Dam and its Types

Dams are barriers that hold water or underground streams. They help in suppressing floods and also help in activities such as cultivation and aquaculture. The topics discussed in the section are of great importance to broaden the existing knowledge on dams.

Dam

A dam is a barrier that impounds water or underground streams. Reservoirs created by dams not only suppress floods but also provide water for activities such as irrigation, human consumption, industrial use, aquaculture, and navigability. Hydropower is often used in conjunction with dams to generate electricity. A dam can also be used to collect water or for storage of water which can be evenly distributed between locations. Dams generally serve the primary purpose of retaining water, while other structures such as floodgates or levees (also known as dikes) are used to manage or prevent water flow into specific land regions.

Karapuzha Dam, an earth dam in the Indian state of Kerala

The word *dam* can be traced back to Middle English, and before that, from Middle Dutch, as seen in the names of many old cities. The first known appearance of *dam* stems from 1165. However, there is one village, Obdam, that is already mentioned in 1120. The word seems to be related to the Greek word *taphos*, meaning "grave" or "grave hill". So the word should be understood as "dike from dug out earth". The names of more than 40 places (with minor changes) from the Middle Dutch era (1150–1500 CE) such as Amsterdam (founded as 'Amstelredam' in the late 12th century) and Rotterdam, also bear testimony to the use of the word in Middle Dutch at that time.

History

Ancient Dams

Early dam building took place in Mesopotamia and the Middle East. Dams were used to control the water level, for Mesopotamia's weather affected the Tigris and Euphrates rivers.

The earliest known dam is the Jawa Dam in Jordan, 100 kilometres (62 mi) northeast of the capital Amman. This gravity dam featured an originally 9-metre-high (30 ft) and 1 m-wide (3.3 ft) stone wall, supported by a 50 m-wide (160 ft) earth rampart. The structure is dated to 3000 BC.

The Ancient Egyptian Sadd-el-Kafara Dam at Wadi Al-Garawi, located about 25 km (16 mi) south of Cairo, was 102 m (335 ft) long at its base and 87 m (285 ft) wide. The structure was built around 2800 or 2600 BC as a diversion dam for flood control, but was destroyed by heavy rain during construction or shortly afterwards. During the Twelfth Dynasty in the 19th century BC, the Pharaohs Senosert III, Amenemhat III and Amenmehat IV dug a canal 16 km (9.9 mi) long linking the Fayum Depression to the Nile in Middle Egypt. Two dams called Ha-Uar running east-west were built to retain water during the annual flood and then release it to surrounding lands. The lake called "Mer-wer" or Lake Moeris covered 1,700 km² (660 sq mi) and is known today as Berkat Qaroun.

One of the engineering wonders of the ancient world was the Great Dam of Marib in Yemen. Initiated somewhere between 1750 and 1700 BC, it was made of packed earth - triangular in cross section, 580 m (1,900 ft) in length and originally 4 m (13 ft) high - running between two groups of rocks on either side, to which it was linked by substantial stonework. Repairs were carried out during various periods, most important around 750 BC, and 250 years later the dam height was increased to 7 m (23 ft). After the end of the Kingdom of Saba, the dam fell under the control of the □imyarites (~115 BC) who undertook further improvements, creating a structure 14 m (46 ft) high, with five spillway channels, two masonry-reinforced sluices, a settling pond, and a 1,000 m (3,300 ft) canal to a distribution tank. These extensive works were not actually finalized until 325 AD and allowed the irrigation of 25,000 acres (100 km²).

By the mid-late 3rd century BC, an intricate water-management system within Dholavira in modern-day India was built. The system included 16 reservoirs, dams and various channels for collecting water and storing it.

Eflatun Pınar is a Hittite dam and spring temple near Konya, Turkey. It is thought to be from the time of the Hittite empire between the 15th and 13th century BC.

The Kallanai is constructed of unhewn stone, over 300 m (980 ft) long, 4.5 m (15 ft) high and 20 m (66 ft) wide, across the main stream of the Kaveri river in Tamil Nadu, South India. The basic structure dates to the 2nd century AD and is considered one of the oldest water-diversion or water-regulator structures in the world which is still in use. The purpose of the dam was to divert the waters of the Kaveri across the fertile delta region for irrigation via canals.

Du Jiang Yan is the oldest surviving irrigation system in China that included a dam that directed waterflow. It was finished in 251 BC. A large earthen dam, made by Sunshu Ao, the prime minister of Chu (state), flooded a valley in modern-day northern Anhui province that created an enormous irrigation reservoir (100 km (62 mi) in circumference), a reservoir that is still present today.

Roman Engineering

Roman dam construction was characterized by "the Romans' ability to plan and organize engineering construction on a grand scale." Roman planners introduced the then-novel concept of large reservoir dams which could secure a permanent water supply for urban settlements over the dry season. Their pioneering use of water-proof hydraulic mortar and particularly Roman concrete allowed for much larger dam structures than previously built, such as the Lake Homs Dam, possibly the largest water barrier to that date, and the Harbaqa Dam, both in Roman Syria. The highest Roman dam was the Subiaco Dam near Rome; its record height of 50 m (160 ft) remained unsurpassed until its accidental destruction in 1305.

The Roman dam at Cornalvo in Spain has been in use for almost two millennia.

Roman engineers made routine use of ancient standard designs like embankment dams and masonry gravity dams. Apart from that, they displayed a high degree of inventiveness, introducing most of the other basic dam designs which had been unknown until then. These include arch-gravity dams, arch dams, buttress dams and multiple arch buttress dams, all of which were known and employed by the 2nd century AD. Roman workforces also were the first to build dam bridges, such as the Bridge of Valerian in Iran.

Remains of the Band-e Kaisar dam, built by the Romans in the 3rd century AD

In Iran, bridge dams such as the Band-e Kaisar were used to provide hydropower through water wheels, which often powered water-raising mechanisms. One of the first was the Roman-built dam bridge in Dezful, which could raise water 50 cubits in height for the water supply to all houses in the town. Also diversion dams were known. Milling dams were introduced which the Muslim engineers called the *Pul-i-Bulaiti*. The first was built at Shustar on the River Karun, Iran, and many of these were later built in other parts of the Islamic world. Water was conducted from the back of the dam through a large pipe to drive a water wheel and watermill. In the 10th century, Al-Muqaddasi described several dams in Persia. He reported that one in Ahwaz was more than 910 m (3,000 ft) long, and that it had many water-wheels raising the water into aqueducts through which it flowed into reservoirs of the city. Another one, the Band-i-Amir dam, provided irrigation for 300 villages.

Middle Ages

In the Netherlands, a low-lying country, dams were often applied to block rivers in order to regulate the water level and to prevent the sea from entering the marsh lands. Such dams often marked the beginning of a town or city because it was easy to cross the river at such a place, and often gave rise to the respective place's names in Dutch.

For instance the Dutch capital Amsterdam (old name *Amstelredam*) started with a dam through the river Amstel in the late 12th century, and Rotterdam started with a dam through the river Rotte, a minor tributary of the Nieuwe Maas. The central square of Amsterdam, covering the original place of the 800-year-old dam, still carries the name *Dam Square* or simply *the Dam*.

Industrial Revolution

An engraving of the Rideau Canal locks at Bytown

The Romans were the first to build arch dams, where the reaction forces from the abutment stabilizes the structure from the external hydrostatic pressure, but it was only in the 19th century that the engineering skills and construction materials available were capable of building the first large-scale arch dams.

Three pioneering arch dams were built around the British Empire in the early 19th century. Henry Russel of the Royal Engineers oversaw the construction of the Mir Alam dam in 1804 to supply water to the city of Hyderabad (it is still in use today). It had a height of 12 m (39 ft) and consisted of 21 arches of variable span.

In the 1820s and 30s, Lieutenant-Colonel John By supervised the construction of the Rideau Canal in Canada near modern-day Ottawa and built a series of curved masonry dams as part of the waterway system. In particular, the Jones Falls Dam, built by John Redpath, was completed in 1832 as the largest dam in North America and an engineering marvel. In order to keep the water in control during construction, two sluices, artificial channels for conducting water, were kept open in the dam. The first was near the base of the dam on its east side. A second sluice was put in on the west side of the dam, about 20 ft (6.1 m) above the base. To make the switch from the lower to upper sluice, the outlet of Sand Lake was blocked off.

Hunts Creek near the city of Parramatta, Australia, was dammed in the 1850s, to cater for the demand for water from the growing population of the city. The masonry arch dam wall was designed by Lieutenant Percy Simpson who was influenced by the advances in dam engineering techniques made by the Royal Engineers in India. The dam cost £17,000 and was completed in 1856 as the first engineered dam built in Australia, and the second arch dam in the world built to mathematical specifications.

Masonry arch wall, Parramatta, New South Wales, the first engineered dam built in Australia

The first such dam was opened two years earlier in France. It was the first French arch dam of the industrial era, and it was built by François Zola in the municipality of Aix-en-Provence to improve the supply of water after the 1832 cholera outbreak devastated the area. After royal approval was granted in 1844, the dam was constructed over the following decade. Its construction was carried out on the basis of the mathematical results of scientific stress analysis.

The 75-miles dam near Warwick, Australia, was possibly the world's first concrete arch dam. Designed by Henry Charles Stanley in 1880 with an overflow spillway and a special water outlet, it was eventually heightened to 10 m (33 ft).

In the latter half of the nineteenth century, significant advances in the scientific theory of masonry dam design were made. This transformed dam design from an art based on empirical methodology to a profession based on a rigorously applied scientific theoretical framework. This new emphasis was centered around the engineering faculties of universities in France and in the United Kingdom. William John Macquorn Rankine at the University of Glasgow pioneered the theoretical understanding of dam structures in his 1857 paper *On the Stability of Loose Earth*. Rankine theory provided a good understanding of the principles behind dam design. In France, J. Augustin Tortene de Sazilly explained the mechanics of vertically faced masonry gravity dams, and Zola's dam was the first to be built on the basis of these principles.

Large Dams

The era of large dams was initiated with the construction of the Aswan Low Dam in Egypt in 1902, a gravity masonry buttress dam on the Nile River. Following their 1882 invasion and occupation of Egypt, the British began construction in 1898. The project was designed by Sir William Willcocks and involved several eminent engineers of the time, including Sir Benjamin Baker and Sir John Aird, whose firm, John Aird & Co., was the main contractor. Capital and financing were furnished by Ernest Cassel. When initially constructed between 1899 and 1902, nothing of its scale had ever been attempted; on completion, it was the largest masonry dam in the world.

The Hoover Dam by Ansel Adams, 1942

The Hoover Dam is a massive concrete arch-gravity dam, constructed in the Black Canyon of the Colorado River, on the border between the US states of Arizona and Nevada between 1931 and 1936 during the Great Depression. In 1928, Congress authorized the project to build a dam that would control floods, provide irrigation water and produce hydroelectric power. The winning bid to build the dam was submitted by a consortium called Six Companies, Inc. Such a large concrete structure had never been built before, and some of the techniques were unproven. The torrid summer weather and the lack of facilities near the site also presented difficulties. Nevertheless, Six Companies turned over the dam to the federal government on 1 March 1936, more than two years ahead of schedule.

By 1997, there were an estimated 800,000 dams worldwide, some 40,000 of them over 15 m (49 ft) high. In 2014, scholars from the University of Oxford published a study of the cost of large dams – based on the largest existing dataset – documenting significant cost overruns for a majority of dams and questioning whether benefits typically offset costs for such dams.

Types of Dams

Dams can be formed by human agency, natural causes, or even by the intervention of wildlife such as beavers. Man-made dams are typically classified according to their size (height), intended purpose or structure.

By Structure

Based on structure and material used, dams are classified as easily created without materials, arch-gravity dams, embankment dams or masonry dams, with several subtypes.

Arch Dams

In the arch dam, stability is obtained by a combination of arch and gravity action. If the upstream face is vertical the entire weight of the dam must be carried to the foundation by gravity, while the distribution of the normal hydrostatic pressure between vertical cantilever and arch action will depend upon the stiffness of the dam in a vertical and horizontal direction. When the upstream face is sloped the distribution is more complicated. The normal component of the weight of the arch ring may be taken by the arch action, while the normal hydrostatic pressure will be distributed as described above. For this type of dam, firm reliable supports at the abutments (either buttress or canyon side wall) are more important. The most desirable place for an arch dam is a narrow canyon with steep side walls composed of sound rock. The safety of an arch dam is dependent on the strength of the side wall abutments, hence not only should the arch be well seated on the side walls but also the character of the rock should be carefully inspected.

Gordon Dam, Tasmania, is an arch dam.

Daniel-Johnson Dam, Quebec, is a multiple-arch buttress dam.

Two types of single-arch dams are in use, namely the constant-angle and the constant-radius dam. The constant-radius type employs the same face radius at all elevations of the dam, which means that as the channel grows narrower towards the bottom of the dam the central angle subtended by the face of the dam becomes smaller. Jones Falls Dam, in Canada, is a constant radius dam. In a constant-angle dam, also known as a variable radius dam, this subtended angle is kept a constant and the variation in distance between the abutments at various levels are taken care of by varying the radii. Constant-radius dams are much less common than constant-angle dams. Parker Dam on the Colorado River is a constant-angle arch dam.

A similar type is the double-curvature or thin-shell dam. Wildhorse Dam near Mountain City, Nevada, in the United States is an example of the type. This method of construction minimizes the amount of concrete necessary for construction but transmits large loads to the foundation and abutments. The appearance is similar to a single-arch dam but with a distinct vertical curvature to it as well lending it the vague appearance of a concave lens as viewed from downstream.

The multiple-arch dam consists of a number of single-arch dams with concrete buttresses as the supporting abutments, as for example the Daniel-Johnson Dam, Québec, Canada. The multiple-arch dam does not require as many buttresses as the hollow gravity type, but requires good rock foundation because the buttress loads are heavy.

Gravity Dams

The Grand Coulee Dam is an example of a solid gravity dam.

In a gravity dam, the force that holds the dam in place against the push from the water is Earth's gravity pulling down on the mass of the dam. The water presses laterally (downstream) on the dam, tending to overturn the dam by rotating about its toe (a point at the bottom downstream side of the dam). The dam's weight counteracts that force, tending to rotate the dam the other way about its toe. The designer ensures that the dam is heavy enough that the dam's weight wins that contest. In engineering terms, that is true whenever the resultant of the forces of gravity acting on the dam and water pressure on the dam acts in a line that passes upstream of the toe of the dam.

Furthermore, the designer tries to shape the dam so if one were to consider the part of dam above any particular height to be a whole dam itself, that dam also would be held in place by gravity. i.e. there is no tension in the upstream face of the dam holding the top of the dam down. The designer does this because it is usually more practical to make a dam of material essentially just piled up than to make the material stick together against vertical tension.

Note that the shape that prevents tension in the upstream face also eliminates a balancing compression stress in the downstream face, providing additional economy.

For this type of dam, it is essential to have an impervious foundation with high bearing strength.

When situated on a suitable site, a gravity dam can prove to be a better alternative to other types of dams. When built on a carefully studied foundation, the gravity dam probably represents the best developed example of dam building. Since the fear of flood is a strong motivator in many regions, gravity dams are being built in some instances where an arch dam would have been more economical.

Gravity dams are classified as "solid" or "hollow" and are generally made of either concrete or masonry. The solid form is the more widely used of the two, though the hollow dam is frequently more economical to construct. Grand Coulee Dam is a solid gravity dam and Braddock Locks & Dam is a hollow gravity dam.

Arch-gravity Dams

The Hoover Dam is an example of an arch-gravity dam.

A gravity dam can be combined with an arch dam into an arch-gravity dam for areas with massive amounts of water flow but less material available for a purely gravity dam. The inward compression of the dam by the water reduces the lateral (horizontal) force acting on the dam. Thus, the gravitation force required by the dam is lessened, i.e. the dam does not need to be so massive. This enables thinner dams and saves resources.

Barrages

The Koshi Barrage

A barrage dam is a special kind of dam which consists of a line of large gates that can be opened or closed to control the amount of water passing the dam. The gates are set between flanking piers which are responsible for supporting the water load, and are often used to control and stabilize water flow for irrigation systems. An example of this type of dam is the now-decommissioned Red Bluff Diversion Dam on the Sacramento River near Red Bluff, California.

Barrages that are built at the mouths of rivers or lagoons to prevent tidal incursions or utilize the tidal flow for tidal power are known as tidal barrages.

Embankment Dams

Embankment dams are made from compacted earth, and have two main types, rock-fill and earth-fill dams. Embankment dams rely on their weight to hold back the force of water, like gravity dams made from concrete.

Rock-fill Dams

The Gathright Dam in Virginia is a rock-fill embankment dam.

Rock-fill dams are embankments of compacted free-draining granular earth with an impervious zone. The earth utilized often contains a high percentage of large particles, hence the term "rock-fill". The impervious zone may be on the upstream face and made of masonry, concrete, plastic membrane, steel sheet piles, timber or other material. The impervious zone may also be within the embankment in which case it is referred to as a *core*. In the instances where clay is utilized as the impervious material the dam is referred to as a *composite* dam. To prevent internal erosion of clay into the rock fill due to seepage forces, the core is separated using a filter. Filters are specifically graded soil designed to prevent the migration of fine grain soil particles. When suitable material is at hand, transportation is minimized leading to cost savings during construction. Rock-fill dams are resistant to damage from earthquakes. However, inadequate quality control during construction can lead to poor compaction and sand in the embankment which can lead to liquefaction of the rock-fill during an earthquake. Liquefaction potential can be reduced by keeping susceptible material from being saturated, and by providing adequate compaction during construction. An example of a rock-fill dam is New Melones Dam in California or the Fierza Dam in Albania.

A core that is growing in popularity is asphalt concrete. The majority of such dams are built with rock and/or gravel as the main fill material. Almost 100 dams of this design have now been built worldwide since the first such dam was completed in 1962. All asphalt-concrete core dams built so far have an excellent performance record. The type of asphalt used is a viscoelastic-plastic material that can adjust to the movements and deformations imposed on the embankment as a whole, and to settlements in the foundation. The flexible properties of the asphalt make such dams especially suited in earthquake regions.

For the Moglicë Hydro Power Plant in Albania the Norwegian power company Statkraft is currently building an asphalt-core rock-fill dam. Upon completion in 2018 the 320 m long, 150 m high and 460 m wide dam is anticipated to be the world's highest of its kind.

Concrete-face Rock-fill Dams

A concrete-face rock-fill dam (CFRD) is a rock-fill dam with concrete slabs on its upstream face. This design provides the concrete slab as an impervious wall to prevent leakage and also a structure without concern for uplift pressure. In addition, the CFRD design is flexible for topography, faster to construct and less costly than earth-fill dams. The CFRD concept originated during the

California Gold Rush in the 1860s when miners constructed rock-fill timber-face dams for sluice operations. The timber was later replaced by concrete as the design was applied to irrigation and power schemes. As CFRD designs grew in height during the 1960s, the fill was compacted and the slab's horizontal and vertical joints were replaced with improved vertical joints. In the last few decades, the design has become popular.

Currently, the tallest CFRD in the world is the 233 m-tall (764 ft) Shuibuya Dam in China which was completed in 2008.

Earth-fill Dams

Earth-fill dams, also called earthen dams, rolled-earth dams or simply earth dams, are constructed as a simple embankment of well compacted earth. A homogeneous rolled-earth dam is entirely constructed of one type of material but may contain a drain layer to collect seep water. A zoned-earth dam has distinct parts or zones of dissimilar material, typically a locally plentiful shell with a watertight clay core. Modern zoned-earth embankments employ filter and drain zones to collect and remove seep water and preserve the integrity of the downstream shell zone. An outdated method of zoned earth dam construction utilized a hydraulic fill to produce a watertight core. Rolled-earth dams may also employ a watertight facing or core in the manner of a rock-fill dam. An interesting type of temporary earth dam occasionally used in high latitudes is the frozen-core dam, in which a coolant is circulated through pipes inside the dam to maintain a watertight region of permafrost within it.

Tarbela Dam is a large dam on the Indus River in Pakistan. It is located about 50 km (31 mi) northwest of Islamabad, and a height of 485 ft (148 m) above the river bed and a reservoir size of 95 sq mi (250 km²) makes it the largest earth-filled dam in the world. The principal element of the project is an embankment 9,000 feet (2,700 m) long with a maximum height of 465 feet (142 m). The total volume of earth and rock used for the project is approximately 200 million cubic yards (152.8 million cu. meters) which makes it one of the largest man-made structures in the world.

Because earthen dams can be constructed from materials found on-site or nearby, they can be very cost-effective in regions where the cost of producing or bringing in concrete would be prohibitive.

By Size

International standards (including the International Commission on Large Dams, ICOLD) define *large dams* as higher than 15 m (49 ft) and *major dams* as over 150 m (490 ft) in height. The *Report of the World Commission on Dams* also includes in the *large* category, dams, such as barrages, which are between 5 and 15 m (16 and 49 ft) high with a reservoir capacity of more than 3 million cubic metres (2,400 acre·ft).

The tallest dam in the world is the 300 m-high (980 ft) Nurek Dam in Tajikistan.

By Use

Saddle Dam

A saddle dam is an auxiliary dam constructed to confine the reservoir created by a primary dam either to permit a higher water elevation and storage or to limit the extent of a reservoir for increased efficiency. An auxiliary dam is constructed in a low spot or "saddle" through which the reservoir would otherwise escape. On occasion, a reservoir is contained by a similar structure called a dike to prevent inundation of nearby land. Dikes are commonly used for reclamation of arable land from a shallow lake. This is similar to a levee, which is a wall or embankment built along a river or stream to protect adjacent land from flooding.

Weir

A weir (also sometimes called an *overflow dam*) is a type of small overflow dam that is often used within a river channel to create an impoundment lake for water abstraction purposes and which can also be used for flow measurement or retardation.

Check Dam

A check dam is a small dam designed to reduce flow velocity and control soil erosion. Conversely, a wing dam is a structure that only partly restricts a waterway, creating a faster channel that resists the accumulation of sediment.

Dry Dam

A dry dam, also known as a flood retarding structure, is a dam designed to control flooding. It normally holds back no water and allows the channel to flow freely, except during periods of intense flow that would otherwise cause flooding downstream.

Diversionary Dam

A diversionary dam is a structure designed to divert all or a portion of the flow of a river from its natural course. The water may be redirected into a canal or tunnel for irrigation and/or hydroelectric power production.

Underground Dam

Underground dams are used to trap groundwater and store all or most of it below the surface for extended use in a localized area. In some cases they are also built to prevent saltwater from intruding into a freshwater aquifer. Underground dams are typically constructed in areas where water resources are minimal and need to be efficiently stored, such as in deserts and on islands like the Fukuzato Dam in Okinawa, Japan. They are most common in northeastern Africa and the arid areas of Brazil while also being used in the southwestern United States, Mexico, India, Germany, Italy, Greece, France and Japan.

There are two types of underground dams: a *sub-surface* and a *sand-storage* dam. A sub-surface dam is built across an aquifer or drainage route from an impervious layer (such as solid bedrock) up to just below the surface. They can be constructed of a variety of materials to include bricks, stones, concrete, steel or PVC. Once built, the water stored behind the dam raises the water table and is then extracted with wells. A sand-storage dam is a weir built in stages across a stream or wadi. It must be strong, as floods will wash over its crest. Over time, sand accumulates in layers behind the dam, which helps store water and, most importantly, prevent evaporation. The stored water can be extracted with a well, through the dam body, or by means of a drain pipe.

Tailings Dam

A tailings dam is typically an earth-fill embankment dam used to store tailings, which are produced during mining operations after separating the valuable fraction from the uneconomic fraction of an ore. Conventional water retention dams can serve this purpose, but due to cost, a tailings dam is more viable. Unlike water retention dams, a tailings dam is raised in succession throughout the life of the particular mine. Typically, a base or starter dam is constructed, and as it fills with a mixture of tailings and water, it is raised. Material used to raise the dam can include the tailings (depending on their size) along with dirt.

There are three raised tailings dam designs, the *upstream*, *downstream* and *centerline*, named according to the movement of the crest during raising. The specific design used is dependent upon topography, geology, climate, the type of tailings, and cost. An upstream tailings dam consists of trapezoidal embankments being constructed on top but toe to crest of another, moving the crest further upstream. This creates a relatively flat downstream side and a jagged upstream side which is supported by tailings slurry in the impoundment. The downstream design refers to the successive raising of the embankment that positions the fill and crest further downstream. A centerlined dam has sequential embankment dams constructed directly on top of another while fill is placed on the downstream side for support and slurry supports the upstream side.

Because tailings dams often store toxic chemicals from the mining process, they have an impervious liner to prevent seepage. Water/slurry levels in the tailings pond must be managed for stability and environmental purposes as well.

By Material

Steel Dams

Red Ridge steel dam, built 1905, Michigan

A steel dam is a type of dam briefly experimented with around the start of the 20th century which uses steel plating (at an angle) and load-bearing beams as the structure. Intended as permanent structures, steel dams were an (arguably failed) experiment to determine if a construction technique could be devised that was cheaper than masonry, concrete or earthworks, but sturdier than timber crib dams.

Timber Dams

A timber crib dam in Michigan, photographed in 1978

Timber dams were widely used in the early part of the industrial revolution and in frontier areas due to ease and speed of construction. Rarely built in modern times because of their relatively short lifespan and the limited height to which they can be built, timber dams must be kept constantly wet in order to maintain their water retention properties and limit deterioration by rot, similar to a barrel. The locations where timber dams are most economical to build are those where timber is plentiful, cement is costly or difficult to transport, and either a low head diversion dam is required or longevity is not an issue. Timber dams were once numerous, especially in the North American West, but most have failed, been hidden under earth embankments, or been replaced with entirely new structures. Two common variations of timber dams were the *crib* and the *plank*.

Timber crib dams were erected of heavy timbers or dressed logs in the manner of a log house and the interior filled with earth or rubble. The heavy crib structure supported the dam's face and the weight of the water. Splash dams were timber crib dams used to help float logs downstream in the late 19th and early 20th centuries.

Timber plank dams were more elegant structures that employed a variety of construction methods utilizing heavy timbers to support a water retaining arrangement of planks.

Other Types

Cofferdams

A cofferdam during the construction of locks at the Montgomery Point Lock and Dam

A cofferdam is a barrier, usually temporary, constructed to exclude water from an area that is normally submerged. Made commonly of wood, concrete, or steel sheet piling, cofferdams are used to allow construction on the foundation of permanent dams, bridges, and similar structures. When the project is completed, the cofferdam will usually be demolished or removed unless the area requires continuous maintenance.

Common uses for cofferdams include construction and repair of offshore oil platforms. In such cases the cofferdam is fabricated from sheet steel and welded into place under water. Air is pumped into the space, displacing the water and allowing a dry work environment below the surface.

Natural Dams

Dams can also be created by natural geological forces. Volcanic dams are formed when lava flows, often basaltic, intercept the path of a stream or lake outlet, resulting in the creation of a natural impoundment. An example would be the eruptions of the Uinkaret volcanic field about 1.8 million–10,000 years ago, which created lava dams on the Colorado River in northern Arizona in the United States. The largest such lake grew to about 800 km (500 mi) in length before the failure of its dam. Glacial activity can also form natural dams, such as the damming of the Clark Fork in Montana by the Cordilleran Ice Sheet, which formed the 7,780 km² (3,000 sq mi) Glacial Lake Missoula near the end of the last Ice Age. Moraine deposits left behind by glaciers can also dam rivers to form lakes, such as at Flathead Lake, also in Montana.

Natural disasters such as earthquakes and landslides frequently create landslide dams in mountainous regions with unstable local geology. Historical examples include the Usoi Dam in Tajikistan, which blocks the Murghab River to create Sarez Lake. At 560 m (1,840 ft) high, it is the tallest dam in the world, including both natural and man-made dams. A more recent example would be the creation of Attabad Lake by a landslide on Pakistan's Hunza River.

Natural dams often pose significant hazards to human settlements and infrastructure. The resulting lakes often flood inhabited areas, while a catastrophic failure of the dam could cause even greater damage, such as the failure of western Wyoming's Gros Ventre landslide dam in 1927, which wiped out the town of Kelly and resulted in the deaths of six people.

Beaver Dams

Beavers create dams primarily out of mud and sticks to flood a particular habitable area. By flooding a parcel of land, beavers can navigate below or near the surface and remain relatively well hidden or protected from predators. The flooded region also allows beavers access to food, especially during the winter.

Construction Elements

Power Generation Plant

As of 2005, hydroelectric power, mostly from dams, supplies some 19% of the world's electricity, and over 63% of renewable energy. Much of this is generated by large dams, although China uses small-scale hydro generation on a wide scale and is responsible for about 50% of world use of this type of power.

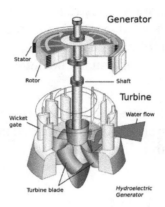

Hydraulic turbine and electric generator

Most hydroelectric power comes from the potential energy of dammed water driving a water turbine and generator; to boost the power generation capabilities of a dam, the water may be run through a large pipe called a penstock before the turbine. A variant on this simple model uses pumped-storage hydroelectricity to produce electricity to match periods of high and low demand, by moving water between reservoirs at different elevations. At times of low electrical demand, excess generation capacity is used to pump water into the higher reservoir. When there is higher demand, water is released back into the lower reservoir through a turbine.

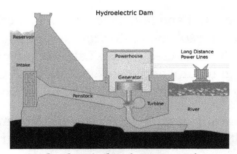

Hydroelectric dam in cross section

Spillways

Spillway on Llyn Brianne dam, Wales, soon after first fill

A spillway is a section of a dam designed to pass water from the upstream side of a dam to the downstream side. Many spillways have floodgates designed to control the flow through the spillway. There are several types of spillway. A *service spillway* or *primary spillway* passes normal flow. An *auxiliary spillway* releases flow in excess of the capacity of the service spillway. An *emergency spillway* is designed for extreme conditions, such as a serious malfunction of the service spillway. A *fuse plug spillway* is a low embankment designed to be overtopped and washed away in the event of a large flood. The elements of a fuse plug are independent free-standing blocks, set side by side which work without any remote control. They allow increasing the normal pool of the dam without compromising the security of the dam because they are designed to be gradually evacuated for exceptional events. They work as fixed weirs at times by allowing over-flow for common floods.

The spillway can be gradually eroded by water flow, including cavitation or turbulence of the water flowing over the spillway, leading to its failure. It was the inadequate design of the spillway which led to the 1889 over-topping of the South Fork Dam in Johnstown, Pennsylvania, resulting in the infamous Johnstown Flood (the "great flood of 1889").

Erosion rates are often monitored, and the risk is ordinarily minimized, by shaping the downstream face of the spillway into a curve that minimizes turbulent flow, such as an ogee curve.

Dam Creation

Common Purposes

Function	Example
Power generation	Hydroelectric power is a major source of electricity in the world. Many countries have rivers with adequate water flow, that can be dammed for power generation purposes. For example, the Itaipu Dam on the Paraná River in South America generates 14 GW and supplied 93% of the energy consumed by Paraguay and 20% of that consumed by Brazil as of 2005.
Water supply	Many urban areas of the world are supplied with water abstracted from rivers pent up behind low dams or weirs. Examples include London, with water from the River Thames, and Chester, with water taken from the River Dee. Other major sources include deep upland reservoirs contained by high dams across deep valleys, such as the Claerwen series of dams and reservoirs.
Stabilize water flow / irrigation	Dams are often used to control and stabilize water flow, often for agricultural purposes and irrigation. Others such as the Berg Strait dam can help to stabilize or restore the water levels of inland lakes and seas, in this case the Aral Sea.
Flood prevention	The Keenleyside Dam on the Columbia River, Canada can store 8.76 km³ (2.10 cu mi) of floodwaters, and the huge Delta Works protects the Netherlands from coastal flooding.
Land reclamation	Dams (often called dykes or levees in this context) are used to prevent ingress of water to an area that would otherwise be submerged, allowing its reclamation for human use.
Water diversion	A typically small dam used to divert water for irrigation, power generation, or other uses, with usually no other function. Occasionally, they are used to divert water to another drainage or reservoir to increase flow there and improve water use in that particular area. dam.
Navigation	Dams create deep reservoirs and can also vary the flow of water downstream. This can in return affect upstream and downstream navigation by altering the river's depth. Deeper water increases or creates freedom of movement for water vessels. Large dams can serve this purpose, but most often weirs and locks are used.

Some of these purposes are conflicting, and the dam operator needs to make dynamic tradeoffs. For example, power generation and water supply would keep the reservoir high, whereas flood prevention would keep it low. Many dams in areas where precipitation fluctuates in an annual cycle will also see the reservoir fluctuate annually in an attempt to balance these difference purposes. Dam management becomes a complex exercise amongst competing stakeholders.

Location

The discharge of Takato Dam

One of the best places for building a dam is a narrow part of a deep river valley; the valley sides then can act as natural walls. The primary function of the dam's structure is to fill the gap in the natural reservoir line left by the stream channel. The sites are usually those where the gap becomes a minimum for the required storage capacity. The most economical arrangement is often a composite structure such as a masonry dam flanked by earth embankments. The current use of the land to be flooded should be dispensable.

Significant other engineering and engineering geology considerations when building a dam include:

- Permeability of the surrounding rock or soil
- Earthquake faults
- Landslides and slope stability
- Water table
- Peak flood flows
- Reservoir silting
- Environmental impacts on river fisheries, forests and wildlife
- Impacts on human habitations
- Compensation for land being flooded as well as population resettlement
- Removal of toxic materials and buildings from the proposed reservoir area

Impact Assessment

Impact is assessed in several ways: the benefits to human society arising from the dam (agriculture, water, damage prevention and power), harm or benefit to nature and wildlife, impact on the geology of an area (whether the change to water flow and levels will increase or decrease stability), and the disruption to human lives (relocation, loss of archeological or cultural matters underwater).

Environmental Impact

Wood and garbage accumulation due to a dam

Reservoirs held behind dams affect many ecological aspects of a river. Rivers topography and dynamics depend on a wide range of flows, whilst rivers below dams often experience long periods of very stable flow conditions or sawtooth flow patterns caused by releases followed by no releases. Water releases from a reservoir including that exiting a turbine usually contain very little suspended sediment, and this in turn can lead to scouring of river beds and loss of riverbanks; for example, the daily cyclic flow variation caused by the Glen Canyon Dam was a contributor to sand bar erosion.

Older dams often lack a fish ladder, which keeps many fish from moving upstream to their natural breeding grounds, causing failure of breeding cycles or blocking of migration paths. Even the presence of a fish ladder does not always prevent a reduction in fish reaching the spawning grounds upstream. In some areas, young fish ("smolt") are transported downstream by barge during parts of the year. Turbine and power-plant designs that have a lower impact upon aquatic life are an active area of research.

A large dam can cause the loss of entire ecospheres, including endangered and undiscovered species in the area, and the replacement of the original environment by a new inland lake.

Large reservoirs formed behind dams have been indicated in the contribution of seismic activity, due to changes in water load and/or the height of the water table.

Dams are also found to have a role in the increase/decrease of global warming. The changing water levels in reservoirs are a source for greenhouse gases like methane. While dams and the water behind them cover only a small portion of earth's surface, they harbour biological activity that can produce large amounts of greenhouse gases.

Human Social Impact

The impact on human society is also significant. Nick Cullather argues in *Hungry World: America's Cold War Battle Against Poverty in Asia* that dam construction requires the state to displace individual people in the name of the common good, and that it often leads to abuses of the masses by planners. He cites Morarji Desai, Interior Minister of India, in 1960 speaking to villagers upset about the Pong Dam, who threatened to "release the waters" and drown the villagers if they did not cooperate.

For example, the Three Gorges Dam on the Yangtze River in China is more than five times the size of the Hoover Dam (U.S.), and creates a reservoir 600 km (370 mi) long to be used for flood control and hydro-power generation. Its construction required the loss of over a million people's homes and their mass relocation, the loss of many valuable archaeological and cultural sites, as well as significant ecological change. During the 2010 China floods, the dam held back a what would have been a disastrous flood and the huge reservoir rose by 4 m (13 ft) overnight.

It is estimated that to date, 40–80 million people worldwide have been physically displaced from their homes as a result of dam construction.

Economics

Construction of a hydroelectric plant requires a long lead time for site studies, hydrological studies, and environmental impact assessments, and are large-scale projects by comparison to traditional power generation based upon fossil fuels. The number of sites that can be economically developed for hydroelectric production is limited; new sites tend to be far from population centers and usually require extensive power transmission lines. Hydroelectric generation can be vulnerable to major changes in the climate, including variations in rainfall, ground and surface water levels, and glacial melt, causing additional expenditure for the extra capacity to ensure sufficient power is available in low-water years.

Once completed, if it is well designed and maintained, a hydroelectric power source is usually comparatively cheap and reliable. It has no fuel and low escape risk, and as an alternative energy source it is cheaper than both nuclear and wind power. It is more easily regulated to store water as needed and generate high power levels on demand compared to wind power.

Dam Failure

The reservoir emptying through the failed Teton Dam

International special sign for works and installations containing dangerous forces

Dam failures are generally catastrophic if the structure is breached or significantly damaged. Routine deformation monitoring and monitoring of seepage from drains in and around larger dams is useful to anticipate any problems and permit remedial action to be taken before structural failure occurs. Most dams incorporate mechanisms to permit the reservoir to be lowered or even drained in the event of such problems. Another solution can be rock grouting – pressure pumping portland cement slurry into weak fractured rock.

During an armed conflict, a dam is to be considered as an "installation containing dangerous forces" due to the massive impact of a possible destruction on the civilian population and the environment. As such, it is protected by the rules of international humanitarian law (IHL) and shall not be made the object of attack if that may cause severe losses among the civilian population. To facilitate the identification, a protective sign consisting of three bright orange circles placed on the same axis is defined by the rules of IHL.

The main causes of dam failure include inadequate spillway capacity, piping through the embankment, foundation or abutments, spillway design error (South Fork Dam), geological instability caused by changes to water levels during filling or poor surveying (Vajont, Malpasset, Testalinden Creek dams), poor maintenance, especially of outlet pipes (Lawn Lake Dam, Val di Stava Dam collapse), extreme rainfall (Shakidor Dam), earthquakes, and human, computer or design error (Buffalo Creek Flood, Dale Dike Reservoir, Taum Sauk pumped storage plant).

A notable case of deliberate dam failure (prior to the above ruling) was the Royal Air Force 'Dambusters' raid on Germany in World War II (codenamed "Operation Chastise"), in which three German dams were selected to be breached in order to damage German infrastructure and manufacturing and power capabilities deriving from the Ruhr and Eder rivers. This raid later became the basis for several films.

Since 2007, the Dutch IJkdijk foundation is developing, with an open innovation model and early warning system for levee/dike failures. As a part of the development effort, full-scale dikes are destroyed in the IJkdijk fieldlab. The destruction process is monitored by sensor networks from an international group of companies and scientific institutions.

Gravity Dam

Willow Creek Dam, a roller-compacted concrete gravity dam

A gravity dam is a dam constructed from concrete or stone masonry and designed to hold back water by primarily utilizing the weight of the material alone to resist the horizontal pressure of water pushing against it. Gravity dams are designed so that each section of the dam is stable, independent of any other dam section.

Gravity dams generally require stiff rock foundations of high bearing strength (slightly weathered to fresh); although they have been built on soil foundations in rare cases. The bearing strength of the foundation limits the allowable position of the resultant which influences the overall stability. Also, the stiff nature of the gravity dam structure is unforgiving to differential foundation settlement, which can induce cracking of the dam structure.

Gravity dams provide some advantages over embankment dams. The main advantage is that they can tolerate minor over-topping flows as the concrete is resistant to scouring. This reduces the requirements for a cofferdam during construction and the sizing of the spillway. Large overtopping flows are still a problem, as they can scour the foundations if not accounted for in the design. A disadvantage of gravity dams is that due to their large footprint, they are susceptible to uplift pressures which act as a de-stabilising force. Uplift pressures (buoyancy) can be reduced by internal and foundation drainage systems which reduces the pressures.

During construction, the setting concrete produces an exothermic reaction. This heat expands the plastic concrete and can take up to several decades to cool. When cooling, the concrete is in a stiff state and is susceptible to cracking. It is the designers' task to ensure this doesn't occur.

The most common classification of gravity dams is by the materials composing the structure:

- Concrete dams include

    - mass concrete dams, made of:

        - conventional concrete: Dworshak Dam, Grand Coulee Dam, Three Gorges Dam

        - Roller-Compacted Concrete (RCC): Willow Creek Dam (Oregon), Upper Stillwater Dam

    - masonry: Pathfinder Dam, Cheesman Dam

    - hollow gravity dams, made of reinforced concrete: Braddock Dam

Composite dams are a combination of concrete and embankment dams. Construction materials of composite dams are the same used for concrete and embankment dams.

Gravity dams can be classified by plan (shape):

- Most gravity dams are straight (Grand Coulee Dam).

- Some masonry and concrete gravity dams have the dam axis curved (Shasta Dam, Cheesman Dam) to add stability through arch action.

Gravity dams can be classified with respect to their structural height:

- Low, up to 100 feet.

- Medium high, between 100 and 300 feet.

- High, over 300 feet.

Arch Dam

The Katse Dam, a 185m high concrete arch dam in Lesotho.

The Morrow Point Dam is a double-curvature arch dam.

An arch dam is a solid dam made of concrete that is curved upstream in plan. The arch dam is designed so that the force of the water against it, known as hydrostatic pressure, presses against the arch, compressing and strengthening the structure as it pushes into its foundation or abutments. An arch dam is most suitable for narrow gorges or canyons with steep walls of stable rock to support the structure and stresses. Since they are thinner than any other dam type, they require much less construction material, making them economical and practical in remote areas.

Classification

In general, arch dams are classified based on the ratio of the base thickness to the structural height (b/h) as:

- Thin, for b/h less than 0.2,

- Medium-thick, for b/h between 0.2 and 0.3, and

- Thick, for b/h ratio over 0.3.

Arch dams classified with respect to their structural height are:

- Low dams up to 100 feet (30 m),

- Medium high dams between 100–300 ft (30–91 m),

- High dams over 300 ft (91 m).

History

Shāh Abbās Arch near Kurit Dam - 14th century

The development of arch dams throughout history began with the Romans in the 1st century BC and after several designs and techniques were developed, relative uniformity was achieved in the 20th century. The first known arch dam, the Glanum Dam, also known as the Vallon de Baume Dam, was built by the Romans in France and it dates back to the 1st century BC. The dam was about 12 metres high and 18 metres in length (39 ft × 59 ft). Its radius was about 14 m (46 ft), and it consisted of two masonry walls. The Romans built it to supply nearby Glanum with water.

The Monte Novo Dam in Portugal was another early arch dam built by the Romans in 300 AD. It was 5.7 metres (19 ft) high and 52 m long (171 ft), with a radius of 19 m (62 ft). The curved ends of the dam met with two winged walls that were later supported by two buttresses. The dam also contained two water outlets to drive mills downstream.

The Dara Dam was another arch dam built by the Romans in which the historian Procopius would write of its design: "This barrier was not built in a straight line, but was bent into the shape of a crescent, so that the curve, by lying against the current of the river, might be able to offer still more resistance to the force of the stream."

The Mongols also built arch dams in modern-day Iran. Their earliest was the Kebar Dam built around 1300, which was 26 m (85 ft) high and 55 m (180 ft) long, and had a radius of 35 m (115 ft). Their second dam was built around 1350 and is called the Kurit Dam. After 4 m (13 ft) was added to the dam in 1850, it became 64 m (210 ft) tall and remained the tallest dam in the world until the early 20th century. The Kurit Dam was of masonry design and built in a very narrow canyon. The canyon was so narrow that its crest length is only 44% of its height. The dam is still erect, even though part of its lower downstream face fell off.

The Elche Dam in Elche, Spain was a post-medieval arch dam built in the 1630s by Joanes del Temple and the first in Europe since the Romans. The dam was 26 metres (85 ft) high and 75 metres (246 ft) long, and had a radius of 62 metres (203 ft). This arch dam also rests on winged walls that served as abutments.

In the 20th century, the world's first variable-radius arch dam was built on the Salmon Creek near Juneau, Alaska. The Salmon Creek Dam's upstream face bulged upstream, which relieved pressure on the stronger, curved lower arches near the abutments. The dam also had a larger toe, which offset pressure on the upstream heel of the dam, which now curved more downstream. The technology and economical benefits of the Salmon Creek Dam allowed for larger and taller dam designs. The dam was, therefore, revolutionary, and similar designs were soon adopted around the world, in particular by the U.S. Bureau of Reclamation.

The Inguri Dam in the Caucasas of Georgia.

Pensacola Dam, completed in the state of Oklahoma in 1940, was considered the longest multiple arch dam in the world. Designed by W. R. Holway, it has 51 arches. and a maximum height of 150 ft (46 m) above the river bed. The total length of the dam and its sections is 6,565 ft (2,001 m) while the multiple-arch section is 4,284 ft (1,306 m) long and its combination with the spillway sections measure 5,145 ft (1,568 m). Each arch in the dam has a clear span of 60 ft (18 m) and each buttress is 24 ft (7.3 m) wide.

Arch dam designs would continue to test new limits and designs such as the double- and multiple-curve. The Swiss engineer Alfred Stucky and the U.S. Bureau of Reclamation would develop a method of weight and stress distribution in the 1960s, and arch dam construction in the United States would see its last surge then with dams like the 143-meter double-curved Morrow Point Dam in Colorado, completed in 1968. By the late 20th century, arch dam design reached a relative uniformity in design around the world. Currently, the tallest and largest arch dam in the world is the 292 metres (958 ft) Xiaowan Dam in China, which was completed in 2010. The longest multiple arch with buttress dam in the world is the Daniel-Johnson Dam in Quebec, Canada. It is 214 meters (702 ft) high and 1,314 meters (4,311 ft) long across its crest. It was completed in 1968 and put in service in 1970. In 2013, China completed an arch dam, Jinping 1 Dam, which is world's tallest at 305 metres (1,001 ft).

Pensacola Dam was one of the last multiple arch types built in the United States. Its NRHP application states that this was because three dams of this type failed: (1) Gem Lake Dam, St. Francis Dam (California), Lake Hodges Dam (California). None of these failures were inherently caused by the multiple arch design.

Design

The design of an arch dam is a very complex process. It starts with an initial dam layout, that is continually improved until the design objectives are achieved within the design criteria.

Loads

The main loads for which an arch dam is designed are:

- Dead load
- Hydrostatic load generated by the reservoir and the tailwater
- Temperature load
- Earthquake load

Other miscellaneous loads that affect a dam include: ice and silt loads, and uplift pressure.

The Idukki Dam in Kerala.

Most often, the arch dam is made of concrete and placed in a "V"-shaped valley. The foundation or abutments for an arch dam must be very stable and proportionate to the concrete. There are two basic designs for an arch dam: *constant-radius dams*, which have constant radius of curvature, and *variable-radius dams*, which have both upstream and downstream curves that systematically decrease in radius below the crest. A dam that is *double-curved* in both its horizontal and vertical planes may be called a dome dam. Arch dams with more than one contiguous arch or plane are described as multiple-arch dams. Early examples include the Roman Esparragalejo Dam with later examples such as the Daniel-Johnson Dam (1968) and Itaipu Dam (1982). However, as a result of the failure of the Gleno Dam shortly after it was constructed in 1923, the construction of new multiple arch dams has become less popular.

Contraction joints are normally placed every 20 m in the arch dam and are later filled with grout after the control cools and cures.

Examples of Arch Dams

El Atazar Dam, near Madrid

- Buchanan Dam (example of multiple-arch type)
- Contra Dam
- Daniel-Johnson Dam
- Deriner Dam
- El Atazar Dam
- Flaming Gorge Dam
- Glen Canyon Dam
- Hartbeespoort Dam
- Idukki Dam
- Inguri Dam
- Karun-3 Dam
- Luzzone Dam
- Mauvoisin Dam
- Pensacola Dam (longest multiple-arch type)
- St. Francis Dam
- Victoria Dam
- Xiluodu Dam
- Mratinje Dam

Check Dam

A steel check dam

A check dam is a small, sometimes temporary, dam constructed across a swale, drainage ditch, or waterway to counteract erosion by reducing water flow velocity. Check dams themselves are not a type of new technology; rather, they are an ancient technique dating all the way back to the second century A.D. Check dams are typically, though not always, implemented as a system of several check dams situated at regular intervals across the area of interest.

Concrete check dams

A common application of check dams is in bioswales, which are artificial drainage channels that are designed to remove silt and pollution from runoff.

Function

A check dam placed in the ditch, swale, or channel interrupts the flow of water and flattens the gradient of the channel, thereby reducing the velocity. In turn, this obstruction induces infiltration rather than eroding the channel. They can be used not only to slow flow velocity but also to distribute flows across a swale to avoid preferential paths and guide flows toward vegetation. Although some sedimentation may result behind the dam, check dams do not primarily function as sediment trapping devices.

Check dams could be designed to create small reservoirs, without possibility of silting. A self desilting design was published in 'Invention Intelligence, August, 1987, which while being permanent, would also remove silt as it is formed, keeping the reservoir capacity maximum. The design envisages an awning, going very near the bottom level, extending to the width of the dam, and embedding into the sides. When freshets occur, the silt is automatically carried over to down stream, keeping the reservoir clear.

Applications

Grade Control Mechanism

Check dams have traditionally been implemented in two main environments: across channel bottoms and on hilly slopes. Check dams are used primarily to control water velocity, conserve soil, and improve land. They are used when other flow-control practices, such as lining the channel or creating bioswales is impractical. Accordingly, they are commonly used in degrading temporary channels, in which permanent stabilization is impractical and infeasible in terms of resource allocation and funding due to the short life period. Or, they are used when construction delays and weather conditions prevent timely installation of other erosion control practices. This is typically seen during the construction process of large-scale permanent dams or erosion control. As such, check dams serve as temporary grade-control mechanisms along waterways until resolute stabilization is established or along permanent swales that need protection prior to installation of a non-erodible lining.

Water Quality Control Mechanism

Many check dams tend to form stream pools. Under low-flow circumstances, water either infiltrates into the ground, evaporates, or seeps through or under the dam. Under high flow - flood - conditions, water flows over or through the structure. Coarse and medium-grained sediment from runoff tends to be deposited behind check dams, while finer grains flow through. Extra nutrients, phosphorus, nitrogen, heavy metals, and floating garbage are also trapped by check dams, increasing their effectiveness as water quality control measures.

Arid Regions

In arid areas, check dams are often built to increase groundwater recharge in a process called managed aquifer recharge. Winter runoff thus can be stored in aquifers, from which the water can be withdrawn during the dry season for irrigation, livestock watering, and even drinking water supply. This is particularly useful for small settlements located far from a large urban center as check dams require less reliance on machinery, funding, or advanced understandings as compared to large-scale dam implementation.

Mountainous Regions

As a strategy to stabilize mountain streams, the construction of check dams has a long tradition in many mountainous regions dating back to the 19th century in Europe. Because of the steep slopes in the mountain region, it may be difficult for large construction machinery to reach mountain streams; therefore, check dams have been implemented in place of large-scale dams. Because the typical high slope causes the flow velocity to move faster, a terraced system of multiple closely spaced check dams is typically necessary to reduce velocity and thereby counteract erosion. Such consolidation check dams, built in terraces, attempt to prevent both headward and downward cutting into channel beds while also stabilizing adjacent hill slopes. They are further used to mitigate flood and debris flow hazards.

Design Considerations

Site

Before installing a check dam, careful inspection of the site must be undertaken. The drainage area should be ten acres or less. The waterway should be on a slope of no more than 50% and should have a minimum depth to bedrock of 2 ft. Check dams are often used in natural or constructed channels or swales. They should never be placed in live streams unless approved by appropriate local, state and/or federal authorities.

Materials

Check dams are made of a variety of materials. Because they are typically used as temporary structures, they are often made of cheap and accessible materials such as rocks, gravel, logs, hay bales, and sandbags. Of these, logs and rock check dams are usually permanent or semi-permanent; and the sandbag check dam is implemented primarily for temporary purposes. Also, there are check dams that are constructed with rockfill or wooden boards. These dams are usually implemented only in small, open channels that drain 10 acres (0.04 km²) or less; and usually do not exceed 2 ft (0.61 m) high. Woven-wire can be used to construct check dams in order to hold fine material in a gully. They are typically utilized in environments where the gully has a moderate slope (less than 10%), small drainage area, and in regions where flood flows do not typically carry large rocks or boulders. In nearly all instances, erosion control blankets, which are biodegradable open-weave blankets, are used in conjunction with check dams. These blankets help enforce vegetation growth on the slopes, shorelines and ditch bottoms.

Size

A check dam should not be more than 2 ft (0.61 m) to 3 ft (0.91 m) high. and the center of the dam should be at least 6 in (0.15 m) lower than its edges. They may kill grass linings in channels if water stays high or sediment load is great. This criteria induces a weir effect, resulting in increased water surface level upstream for some, if not all flow conditions.

Spacing

In order to effectively slow down water velocity to counter the effects of erosion and protect the channel between dams in a larger system, the spacing must be designed properly. The check dams should be spaced such that the toe of the upstream check dam is equal to the elevation of the downstream check dam's crest. By doing so, the water can pond between check dams and thus slow the flow's velocity down substantially as the water progresses downslope.

Advantages

Check dams are a highly effective practice to reduce flow velocities in channels and waterways. Contrasting big dams, check dams have a faster implementation timeline, are cost effective, and are smaller in scope. Because of this, their implementation will not typically displace people or communities nor will they destroy natural resources if careful design considerations are undertaken. Moreover, the dams themselves are simple to construct and do not rely on advanced technologies – thereby they can be applied in more rural and less advanced communities, as they have been in India's drylands for some time now.

Limitations

Check dams still require maintenance and sediment removal practices. They become more difficult to implement on steep slopes, as velocity is higher and thereby the distance between dams must be shortened. Check dams, depending on the material used, can have a limited life span but if implemented correctly can be considered permanent though not encouraged.

Maintenance

Check dams require regular maintenance as they are used primarily as a temporary structure and thereby are not designed to withstand long-term use. Dams should be inspected every week that it is sited in the channel and after every large storm. It is important that rubble, litter, and leaves are removed from the upstream side of the dam. This is typically done when the sediment has reached a height of one-half the original height of the dam.

Further, maintenance is required when removing the check dam altogether. In order to ensure the future flow is not adversely altered, the check dam must be fully removed, including any parts that may have been dislodged and washed downstream or any newly developed bare spots where the check dam once was situated.

Cofferdam

A cofferdam on the Ohio River near Olmsted, Illinois, built for the purpose of constructing the Olmsted Lock and Dam

A cofferdam during the construction of locks at the Montgomery Point Lock and Dam

The Havana Harbor wreckage of the USS Maine surrounded by a cofferdam, on 16 June 1911

A cofferdam (also called a coffer) is a temporary enclosure built within, or in pairs across, a body of water and constructed to allow the enclosed area to be pumped out. This pumping creates a dry work environment for the major work to proceed. Enclosed coffers are commonly used for construction and repair of oil platforms, bridge piers and other support structures built within or over water. These cofferdams are usually welded steel structures, with components consisting of sheet piles, wales, and cross braces. Such structures are typically dismantled after the ultimate work is completed.

Cofferdam Uses

For dam construction, two cofferdams are usually built, one upstream and one downstream of the proposed dam, after an alternative diversion tunnel or channel has been provided for the river flow to bypass the dam foundation area. These cofferdams are typically a conventional embankment dam of both earth- and rock-fill, but concrete or some sheet piling also may be used. Typically, upon completion of the dam and associated structures, the downstream coffer is removed and the upstream coffer is flooded as the diversion is closed and the reservoir begins to fill. Dependent upon the geography of a dam site, in some applications, a "U"-shaped cofferdam is used in the construction of one half of a dam. When complete, the cofferdam is removed and a similar one is created on the opposite side of the river for the construction of the dam's other half.

The cofferdam is also used on occasion in the shipbuilding and ship repair industry, when it is not practical to put a ship in drydock for repair or alteration. An example of such an application is certain ship lengthening operations. In some cases a ship is actually cut in two while still in the water, and a new section of ship is floated in to lengthen the ship. Torch cutting of the hull is done inside a cofferdam attached directly to the hull of the ship; the cofferdam is then detached before the hull sections are floated apart. The cofferdam is later replaced while the hull sections are welded together again. As expensive as this may be to accomplish, use of a drydock may be even more expensive.

A 100-ton open caisson that was lowered more than a mile to the sea floor in attempts to stop the flow of oil in the Deepwater Horizon oil spill has been called a cofferdam. It did not work, as methane hydrates froze in the upper levels preventing the containment.

Naval Architecture

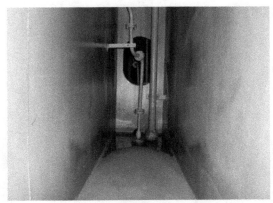

Inside of a cofferdam on a vessel

A cofferdam may also refer to an insulating space between two watertight bulkheads or decks within a ship. A cofferdam may be a void (empty) space or a ballast space. Cofferdams are usually employed to ensure oil or other chemicals do not leak into machinery spaces. If two different cargoes that react dangerously with each other are carried on the same vessel, one or more cofferdams are usually required between the cargo spaces.

Portable Cofferdams

Portable cofferdams are inflatable cofferdams that can be reused. Cofferdams of this type are stretched across the site, then inflated with water from the prospected dry area. Once the area is dry, water still remaining from the dry area can be siphoned over to the wet area.

Detention Dam

A detention dam is a dam built to catch surface runoff and stream water flow in order to regulate the water flow in areas below the dam. Detention dams are commonly used to reduce the damage caused by flooding or to manage the flow rate through a channel. Detention dams can also be constructed to replenish groundwater and trap sediment. Detention dams are one of three classifications of dams which include: storage dams, diversion dams, and detention dams. Storage dams are used to store water for extended lengths of time. The stored water then can be used for irrigation, livestock, municipal water supply, recreation, and hydroelectric power generation. Diversion dams are used to raise the water level in order to redirect the water to the designated location. The diverted water can be used for supplying irrigation systems or reservoirs.

Purposes

Detention dams are constructed for multiple purposes including: flood control, flow rate restriction, subsurface water replenishment, and sediment trapping. Detention dams are used in flood prone areas for flood control. A detention dam is built at an elevation above the flood prone zone. Flood water is collected in the basin above the dam and released at a flow rate the flood zone and channel can accommodate. Channels can include dikes, canals, streams, drain pipes, and rivers. The basin above a flood detention dam should remain at the lowest water level in order to prevent overtopping. Overtopping is when the water level behind the dam exceeds the dam crest height. The dam crest is the top edge of the dam. Overtopping is caused by extreme flooding or severe waves. The severe waves can be a result of high winds, landslides, and earthquakes. The detention dam design must take into account the probability of overtopping occurring and be designed accordingly. Detention dams built to restrict flow rate regulate the amount water released into channels. Detentions dams used to replenishment the subsurface water or groundwater hold surface runoff to allow the ground to absorb the water. Groundwater is water that has drained into an aquifer due to the force of gravity. An aquifer is layers of permeable soils and rocks below the earth's surface that allow water to accumulated between the rocks and soils. Debris dams are a type of detention dam used to collect sediment to prevent it from flowing into areas where large sediment buildup may be damaging.

Design

Detention dams have two basic designs. The detention dam can be made from concrete or masonry which usually involves the use of a metal reinforcing substructure or frame. The concrete or masonry style dam commonly has a cross sectional shape similar to a right triangle with the sloping face pointing downstream and the perpendicular face pointing upstream. Detention dams can also be made from rock or earth to form a gravity embankment style dam. The cross sectional shape of an earth and rock gravity embankment style dam closely resembles an equal lateral triangle with the angled sides facing upstream and downstream. Detentions dams built today are designed and constructed with large safety factors in order to take into account and compensate for the probability of failure.

Flood Detention Dam Design

Flood detention dams are commonly used as part of a flood or storm water detention system. Flood detention systems combine the use of detention basins, detention dams, and channels in order efficiently collect and regulate the runoff. The detention facilities not only regulate the amount of water that is released, but control the quality of the water that is released. The runoff is monitored in order to prevent harmful contaminates and debris from damaging resources like: lakes, rivers, and wetlands. Flood detention dams are constructed methodically. The watershed topographical data, hydrological records, and geological structure for the area are analyzed in order to determine the most effective location for the flood detention dam(s). The analyzed watershed topographical data, hydrological records, and geological structure display the potential storage capacity, environmental impacts, and physical limitations of the area. As a result, models can be generated to simulate the effectiveness of possible flood detention dam locations and designs. The general design for a flood detention dam has a cross sectional shape of a trapezoid where the longer of the parallel sides is the base of the dam, and the angled sides face upstream and downstream. The flood detention dam has an opening at the top in order to release the flood water at a controlled rate that the channels below can accommodate. The flood detention dam models can be used to determine the necessary dam height and overflow opening size in order to prevent overtopping.

Disadvantages

Detention dams can cause injury and damage if they are not built and maintained correctly. Poorly maintained and older detention dams can pose a reliability threat because they may not meet the current structural safety and hydraulic requirements. For example, a detention dam in a populated area that does conform to the current structural safety and hydraulic requirements has a high probability of failing. If severe flood were to occur, the nonconforming detention dam could be overtopped and breached resulting in injury and damage of the surrounding populated area below the detention dam.

Examples

Sediment Detention Dam Example

The area of Fifteenmile Creek, Wyoming was flagged by the United States Department of the Interior Bureau of Land Management in the 1960s because the area was in need of an aggressive sediment control system. Over the course of 10 years, US$2 million was spent to construct 34 sediment detention dams, 110 reservoirs, and 21 spreader dikes in order to manage the sediment issue. The purpose of the sediment control system was to reduce the amount of suspended sediment in the Bighorn River. The high sediment concentration in the Bighorn River was largely attributed to the drainage from the Fifteenmile Creek erosion. The control system was intended to reduce the sediment amount by 25%. However, 20 years after installation the control system was analyzed and uncovered that the detention dams had been improperly maintained, resulting in sediment detention failure. In addition to poor maintenance, the functional lives of the detention dams were shortened because of the location and climate. As a result, people who live downstream of the Bighorn River have to cope with the effects of the sediment. The high amounts of sediment released from the failed detention dams and control system have increased the cost to filter municipal water due to suspended sediment in the Bighorn River. The high sediment deposits have also damaged fisheries and reduced the amount of water that can be stored in a downstream reservoir.

Flood Detention Dam Example

In San Antonio, Texas, the Olmos Creek detention dam was constructed as a flood detention dam. Even though the Olmos Creek detention dam's primary purpose was as a flood detention dam, the dam also acts as debris or sediment detention dam to trap pollutants from entering regional the water supply. The Olmos Creek detention dam is unique because it is located in an urban area with a large floodplain and the area around the detention dam is used as a recreational and wildlife area. As a result, the Olmos Creek detention dam is a multipurpose facility that can handle large floods, trap pollutants, and provide a recreational and wildlife area for the community.

Diversion Dam

The Faraday Diversion Dam, Clackamas River. This dam slows a normally fast and shallow river for partial diversion to a hydroelectric dam. The diversion tunnel opening can be seen in the upper left.

A diversion dam is a dam that diverts all or a portion of the flow of a river from its natural course. Diversion dams do not generally impound water in a reservoir. Instead, the water is diverted into an artificial water course or canal, which may be used for irrigation or return to the river after passing through hydroelectric generators, flow into a different river or be itself dammed forming a reservoir.

An early diversion dam is the Ancient Egyptian Sadd el-Kafara Dam at Wadi Al-Garawi, which was located about twenty five kilometres south of Cairo. Built around 2600 BC for flood control, the structure was 102 metres long at its base and eighty seven metres wide. It was destroyed by a flood while it was still under construction.

The Imperial Dam diverting the Colorado River in the southwestern United States.

Classification

Diversion dams are one of three classifications of dams which include: storage dams, detention dams, and diversion dams. Storage dams are used to store water for extended lengths of time. The stored water then can be used for irrigation, livestock, municipal water supply, recreation, and hydroelectric power generation. Detention dams are built to catch surface runoff to prevent floods and trap sediment by regulating the flow rate of the runoff into channels downstream. Diversion dams are used to raise the water level in order to redirect the water to the designated location. The diverted water can be used for supplying irrigation systems or reservoirs.

Purpose

Diversion dams are installed to raise the water level of a body of water to allow the water to be redirected. The redirected water can be used to supply irrigation systems, reservoirs, or hydroelectric power generation facilities. The water diverted by the diversion dam to the reservoirs can be used for industrial applications or for municipal water supply.

Construction

The design a diversion dam will fall into one of four basic types: embankment style dams, buttress style dams, arch style dams, and gravity styles dams.

Embankment Style Diversion Dam

Embankment style diversion dams are constructed to counteract the force of the water pushing on the dam by building a dam with enough weight to withstand the force. Embankment dams are commonly made from materials in the surrounding area where the dam is being built. The materials generally include: sand, gravel, and rocks. The combination of these building materials with either clay or an impervious membrane gives the embankment dam its integrity. As a result, the combination of its simple construction and locally available building materials the cost of building an embankment dam is lower than the other types of dams.

Buttress Style Diversion Dam

Buttress style diversion dams are designed using angle supports on the downstream side of the dam. The supports are fixed to the wall of the dam in order to help counteract the force of the water on the dam. Buttress style dams are built across wide valleys that do not have a solid bedrock foundation. Bedrock is solid rock that makes up the upper part of the earth's crust. Bedrock can be made from sedimentary, igneous, and metaphoric rock origins. Buttress dams require extensive steel framework and labor. As a result, buttress style dams are expensive to construct and are seldom built today.

Arch Style Diversion Dam

Arch style diversion dams are designed using an arch shape with the top of the arch facing upstream. The arch shape provides extra strength to counteract the force of the water. Arch style dams are generally constructed in narrow canyons. Arch style dams are commonly made from concrete. To ensure the dam's integrity, a solid contact between the bedrock foundation and the dam's concrete base is required. The dome style dam is a type of arch dam. The dome style dam curves in both the horizontal plane and vertical plane. The arch style dam only curves in the horizontal.

Gravity Style Diversion Dam

Gravity style diversion dams are built to non counteract the force of the water pushing on the dam by building a dam with enough weight to withstand the force. Gravity dams are commonly constructed using masonry or cement. The foundations of the gravity dams are generally constructed on top of a solid bedrock foundation. However, gravity dams can be built over unconsolidated ground as long as proper measures are put in place to stop the flow of water under the dam. If water were to get under the dam, the dam could fail.

Embankment Dam

The Mica Dam in Canada.

Tataragi Dam in Asago, Hyōgo Pref., Japan.

Tarbela Dam in Pakistan.

An embankment dam is a large artificial dam. It is typically created by the placement and compaction of a complex semi-plastic mound of various compositions of soil, sand, clay and/or rock. It has a semi-pervious waterproof natural covering for its surface and a dense, impervious core. This makes such a dam impervious to surface or seepage erosion. Such a dam is composed of fragmented independent material particles. The friction and interaction of particles binds the particles together into a stable mass rather than by the use of a cementing substance.

Types

Embankment dams come in two types: the earth-filled dam (also called an earthen dam or terrain dam) made of compacted earth, and the rock-filled dam. A cross-section of an embankment dam shows a shape like a bank, or hill. Most have a central section or core composed of an impermeable material to stop water from seeping through the dam. The core can be of clay, concrete, or asphalt concrete. This dam type is a good choice for sites with wide valleys. They can be built on hard rock or softer soils. For a rock-fill dam, rock-fill is blasted using explosives to break the rock. Additionally, the rock pieces may need to be crushed into smaller grades to get the right range of size for use in an embankment dam.

Safety

The building of a dam and the filling of the reservoir behind it places a new weight on the floor and sides of a valley. The stress of the water increases linearly with its depth. Water also pushes against the upstream face of the dam, a nonrigid structure that under stress behaves semiplastically, and causes greater need for adjustment (flexibility) near the base of the dam than at shallower water levels. Thus the stress level of the dam must be calculated in advance of building to ensure that its break level threshold is not exceeded.

Overtopping or overflow of an embankment dam beyond its spillway capacity will cause its eventual failure. The erosion of the dam's material by overtopping runoff will remove masses of material whose weight holds the dam in place and against the hydraulic forces acting to move the dam. Even a small sustained overtopping flow can remove thousands of tons of overburden soil from the mass of the dam within hours. The removal of this mass unbalances the forces that stabilize the dam against its reservoir as the mass of water still impounded behind the dam presses against the lightened mass of the embankment, made lighter by surface erosion. As the mass of the dam erodes, the force exerted by the reservoir begins to move the entire structure. The embankment, having almost no elastic strength, would begin to break into separate pieces, allowing the impounded reservoir water to flow between them, eroding and removing even more material as it passes through. In the final stages of failure the remaining pieces of the embankment would offer almost no resistance to the flow of the water and continue to fracture into smaller and smaller sections of earth and/or rock until these would disintegrates into a thick mud soup of earth, rocks and water.

Therefore, safety requirements for the spillway are high, and require it to be capable of containing a maximum flood stage. It is common for its specifications to be written such that it can contain a five hundred year flood. Recently a number of embankment dam overtopping protection systems have been developed. These techniques include the concrete overtopping protection systems, timber cribs, sheet-piles, riprap and gabions, reinforced earth, minimum energy loss weirs, embankment overflow stepped spillways and the precast concrete block protection systems.

Steel Dam

Redridge Steel Dam (upstream side) with a low water level

A steel dam is a type of dam (a structure to impound or retard the flow of water) that is made of steel, rather than the more common masonry, earthworks, concrete or timber construction materials.

Relatively few examples were ever built. Of the three built in the US, two remain, the Ashfork-Bainbridge Steel Dam, built in 1898 in the Arizona desert to supply locomotive water to the Atchison, Topeka and Santa Fe Railway (ATSF), and the Redridge Steel Dam, built 1901, in the Upper Peninsula of Michigan to supply water to stamp mills. The third, the Hauser Lake Dam in Montana, was finished in 1907 but failed in 1908.

Steel dams were found uneconomical after the World War I as the steel prices raised many times compared to cement prices though they are equally sound like other dam building materials. However, their economics are highly favourable in 21st century due to higher onsite labour costs, costly bulk material transportation, availability of more construction time in a year, flexibility in construction plan complying statuary requirements, etc.

Principles of Operation

Steel dams use a series of footings anchored in the earth. These footings hold struts which in turn hold up a series of deck girders which in turn hold steel plates. It is these plates that the water comes in contact with. The girders and plates are angled in the downstream direction so that part of the weight of the water acts with a downward force on the struts and footings, holding them in place. (Consider that, if the plates were vertical, as in a steel cofferdam, all the force would be horizontal and much more massive struts and anchors would be required to counteract the horizontal force and bending moment.)

Direct Strutted

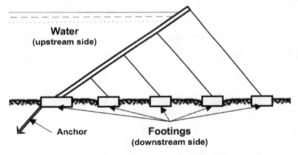

Cross section of a steel dam with direct struts

In the direct strutted version, shown in the illustration at left, all the struts are parallel. There is thus no tensile force in the plate girders.

Cantilever Strutted

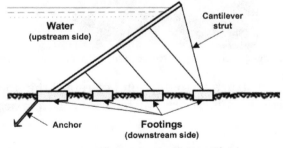

Cross section of a steel dam with cantilever struts

In the cantilever strutted version, shown in the illustration at left, the top strut (or struts, depending on design) can be fashioned into a cantilever truss. By all going to the same footing, the upper part of the deck girders are thus in tension and the moment of the cantilever section is offset by the moment of the water impinging on that section.

Scalloping

In both types of construction, it is typical for the plates to have a scalloped appearance, as can be seen in the Redridge dam illustration, above. It is not known if this is due to the steel stretching or if this was a designed-in feature. It may have been to allow for expansion/contraction as the temperature changed.

Design Tradeoffs

There are two design trade-offs, the girder plate angle and the strut angle. Increasing the girder/plate angle towards the horizontal, the normal component of the force will increase towards vertical; this means that footings do not need to resist as much horizontal force, but requires more steel for a given upstream head. Increasing the strut angle towards vertical reduces the horizontal moment on the footings, reducing the risk of sliding.

Spillways and Pipes

Steel dams may or may not have a spillway. The Ashfork-Bainbridge did not have one but was designed to allow water to pour directly over the crest, while the Redridge had both a spillway and a water pipe to supply water to downstream stamp mills.

Advantages and Disadvantages

Steel Dam proponents claimed some advantages:

- Steel fabrication techniques, even at the turn of the 19th century, allowed for faster and cheaper construction than masonry

- The structure is statically determinate allowing precise calculations of load and member strength needed

- Since steel is more flexible than concrete, they are more resistant to catastrophic failure due to ground settling

- Frost does not affect them the way it does concrete or masonry

- Non–catastrophic leaks can be addressed by welding

- There were also some known disadvantages:

- Constructing good footings is key to a successful dam as they must bear weight, not settle too much and resist horizontal travel.

- The long term strength of the dam is not known. The two examples in the US still standing are not currently under significant water load

- The lightness of the structure means it is more vulnerable to wear due to water vibrations than more massive dams

- Maintenance needs are higher, rust and corrosion must be addressed

- Stresses can be quite concentrated, which could cause stress cracking as a failure mode.

- As with other dams, undermining is a possible failure mode (this is believed to be why the Hauser Lake dam failed.)

References

- Singh, Vijay P.; Ram Narayan Yadava (2003). Water Resources System Operation: Proceedings of the International Conference on Water and Environment. Allied Publishers. p. 508. ISBN 81-7764-548-X. Retrieved 2015-11-09.

- Donald Routledge Hill (1996), "Engineering", p. 759, in Rashed, Roshdi; Morelon, Régis (1996). Encyclopedia of the History of Arabic Science. Routledge. pp. 751–795. ISBN 0-415-12410-7.

- Neves, edited by E. Maranha das (1991). Advances in rockfill structures. Dordrecht: Kluwer Academic. p. 341. ISBN 0-7923-1267-8. Retrieved 2015-11-09.

- Blight, Geoffrey E. (1998). "Construction of Tailings Dams". Case studies on tailings management. Paris, France: International Council on Metals and the Environment. pp. 9–10. ISBN 1-895720-29-X. Retrieved 10 August 2011.

- Herzog, Max A. M. (1999). Practical Dam Analysis. London: Thomas Telford Publishing. pp. 115, 119–126. ISBN 3-8041-2070-9.

- Marsh, William M. (2010). Landscape Planning: Environmental Applications (5th ed.). Danvers, MA: John Wiley & Sons, Inc. pp. 267–268. ISBN 978-0-470-57081-4.

- Melbourne Water (2005). Water Sensitive Urban Design Engineering Procedures: Stormwater. Australia: CSIRO Publishing. p. 140. ISBN 0-643-09092-4. Retrieved 28 October 2014.

- Garcia, Carmelo & Mario Lenzi (2010). Check Dams, Morphological Adjustments and Erosion Control in Torrential Streams. New York: Nova Science Publishers. ISBN 978-1-61761-749-2.

- Cleveland, Cutler J.; Morris, Christopher, eds. (2009). Dictionary of Energy (Expanded ed.). Oxford: Elsevier. p. 133. ISBN 978-0080964911.

- Goings, David B. (2004). K. Lee Lerner; Brenda Wilmoth Lerner, eds. The Gale Encyclopedia of Science (Web) (3rd ed.). Detroit: Gale. pp. 1149–1142. ISBN 978-0787675547.

- Wallace, Jonathan (2005). Carl Mitcham, ed. Encyclopedia of Science, Technology, and Ethics. Detroit: Macmillan Reference USA. pp. 463–465. ISBN 978-0028658315.

- Singh, Vijay P.; Jain, Sharad K.; Tyagi, Aditya (2007). Risk and Reliability Analysis - A Handbook for Civil and Environmental Engineers. American Society of Civil Engineers (ASCE). p. 672. ISBN 978-0784408919.

Various Types of Hydroelectricity Dam

The Three Gorges Dam in China PR is the world's largest power station. Apart from producing electricity, the dam also helps in increasing the Yangtze River's shipping capacity. The other types of hydroelectricity dams are W.A.C. Bennett dam, Itaipu dam, Tarbela dam, Atatürk dam and Hoover dam. This chapter elucidates the main types of hydroelectricity dams.

Three Gorges Dam

The Three Gorges Dam is a hydroelectric dam that spans the Yangtze River by the town of Sandouping, located in Yiling District, Yichang, Hubei province, China. The Three Gorges Dam is the world's largest power station in terms of installed capacity (22,500 MW). In 2014 the dam generated 98.8 TWh of electricity, setting a new world record by 0.17 TWh previously held by the Itaipú Dam on the Brazil/Paraguay border in 2013 of 98.63. But in 2015, the Itaipu power plant resumed the lead in annual worldwide production, producing 89.5 TWh, while production of Three Gorges was 87 TWh.

Except for a ship lift, the dam project was completed and fully functional as of July 4, 2012, when the last of the main water turbines in the underground plant began production. The ship lift was complete in December 2015. Each main water turbine has a capacity of 700 MW. The dam body was completed in 2006. Coupling the dam's 32 main turbines with two smaller generators (50 MW each) to power the plant itself, the total electric generating capacity of the dam is 22,500 MW.

As well as producing electricity, the dam is intended to increase the Yangtze River's shipping capacity and reduce the potential for floods downstream by providing flood storage space. The Chinese government regards the project as a historic engineering, social and economic success, with the design of state-of-the-art large turbines, and a move toward limiting greenhouse gas emissions. However, the dam flooded archaeological and cultural sites and displaced some 1.3 million people, and is causing significant ecological changes, including an increased risk of landslides. The dam has been a controversial topic both domestically and abroad.

History

A large dam across the Yangtze River was originally envisioned by Sun Yat-sen in *The International Development of China*, in 1919. He stated that a dam capable of generating 30 million horsepower (22 GW) was possible downstream of the Three Gorges. In 1932, the Nationalist government, led by Chiang Kai-shek, began preliminary work on plans in the Three Gorges. In 1939, Japanese military forces occupied Yichang and surveyed the area. A design, the Otani plan, was completed for the dam in anticipation of a Japanese victory over China.

In his poem "Swimming" (1956), engraved on the 1954 Flood Memorial in Wuhan, Mao Zedong envisions "walls of stone" to be erected upstream.

In 1944, the United States Bureau of Reclamation chief design engineer, John L. Savage, surveyed the area and drew up a dam proposal for the 'Yangtze River Project'. Some 54 Chinese engineers went to the U.S. for training. The original plans called for the dam to employ a unique method for moving ships; the ships would move into locks located at the lower and upper ends of the dam and then cranes with cables would move the ships from one lock to the next. In the case of smaller water craft, groups of craft would be lifted together for efficiency. It is not known whether this solution was considered for its water-saving performance or because the engineers thought the difference in height between the river above and below the dam too great for alternative methods. Some exploration, survey, economic study, and design work was done, but the government, in the midst of the Chinese Civil War, halted work in 1947.

After the 1949 Communist takeover, Mao Zedong supported the project, but began the Gezhouba Dam project nearby first, and economic problems including the Great Leap Forward and the Cultural Revolution slowed progress. After the 1954 Yangtze River Floods, in 1956, Mao Zedong authored "Swimming", a poem about his fascination with a dam on the Yangtze River. In 1958, after the Hundred Flowers Campaign, some engineers who spoke out against the project were imprisoned.

During the 1980s, the idea of a dam reemerged. The National People's Congress approved the dam in 1992: out of 2,633 delegates, 1,767 voted in favour, 177 voted against, 664 abstained, and 25 members did not vote. Construction started on December 14, 1994. The dam was expected to be fully operational in 2009, but additional projects, such as the underground power plant with six additional generators, delayed full operation until May 2012. The ship lift was completed in 2015. The dam had raised the water level in the reservoir to 172.5 m (566 ft) above sea level by the end of 2008 and the designed maximum level of 175 m (574 ft) by October 2010.

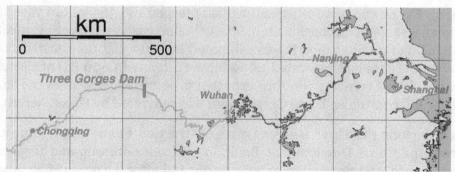

Map of the location of the Three Gorges Dam and the most important cities along the Yangtze River

Composition and Dimensions

Model of the Three Gorges Dam looking upstream, showing the dam body (middle left), the spillway (middle of the dam body) and the ship lift (to the right).

Model of the Three Gorges Dam showing the ship lift and the ship lock. The ship lift is to the right of the dam body with its own designated waterway. The ship locks are to the right (northeast) of the ship lift.

Made of concrete and steel, the dam is 2,335 m (7,661 ft) long and the top of the dam is 185 m (607 ft) above sea level. The project used 27.2×10^6 m³ (35.6×10^6 cu yd) of concrete (mainly for the dam wall), used 463,000 T of steel (enough to build 63 Eiffel Towers), and moved about 102.6×10^6 m³ (134.2×10^6 cu yd) of earth. The concrete dam wall is 181 m (594 ft) high above the rock basis.

When the water level is at its maximum of 175 m (574 ft) above sea level, which is 110 m (361 ft) higher than the river level downstream, the dam reservoir is on average about 660 km (410 mi) in length and 1.12 km (3,675 ft) in width. It contains 39.3 km³ (31,900,000 acre·ft) of water and has a total surface area of 1,045 km² (403 sq mi). On completion, the reservoir flooded a total area of 632 km² (244 sq mi) of land, compared to the 1,350 km² (520 sq mi) of reservoir created by the Itaipu Dam.

Economics

The government estimated that the Three Gorges Dam project would cost 180 billion yuan

(US$22.5 billion). By the end of 2008, spending had reached 148.365 billion yuan, among which 64.613 billion yuan was spent on construction, 68.557 billion yuan on relocating affected residents, and 15.195 billion yuan on financing. It was estimated in 2009 that the construction cost would be recovered when the dam had generated 1,000 terawatt-hours (3,600 PJ) of electricity, yielding 250 billion yuan. Full cost recovery was thus expected to occur ten years after the dam started full operation, but the full cost of the Three Gorges Dam was recovered by December 20, 2013.

Funding sources include the Three Gorges Dam Construction Fund, profits from the Gezhouba Dam, loans from the China Development Bank, loans from domestic and foreign commercial banks, corporate bonds, and revenue from both before and after the dam is fully operational. Additional charges were assessed as follows: Every province receiving power from the Three Gorges Dam had to pay ¥7.00 per MWh extra. Other provinces had to pay an additional charge of ¥4.00 per MWh. The Tibet Autonomous Region pays no surcharge.

Panorama of the Three Gorges Dam

Power Generation and Distribution

Generating Capacity

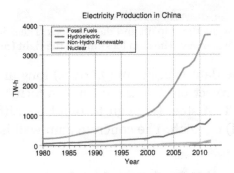

Electricity production in China by source. Compare: The fully completed Three Gorges dam will contribute about 100 TWh of generation per year.

◼thermofossil

◼hydroelectric

◻nuclear

Power generation is managed by China Yangtze Power, a listed subsidiary of China Three Gorges Corporation (CTGC) — a Central Enterprise SOE administered by SASAC. The Three Gorges Dam is the world's largest capacity hydroelectric power station with 34 generators: 32 main generators, each with a capacity of 700 MW, and two plant power generators, each with capacity of 50 MW, making a total capacity of 22,500 MW. Among those 32 main generators, 14 are installed in the north side of the dam, 12 in the south side, and the remaining six in the underground power plant in the mountain south of the dam. The expected annual electricity generation will be over 100 TWh.

Generators

The main generators weigh about 6,000 tonnes each and are designed to produce more than 700 MW of power. The designed head of the generator is 80.6 meters (264 ft). The flow rate varies between 600–950 cubic metres per second (21,000–34,000 cu ft/s) depending on the head available. The greater the head, the less water needed to reach full power. Three Gorges uses Francis turbines. Turbine diameter is 9.7/10.4 m (VGS design/Alstom's design) and rotation speed is 75 revolutions per minute. Rated power is 778 MVA, with a maximum of 840 MVA and a power factor of 0.9. The generator produces electrical power at 20 kV. The outer diameter of the generator stator is 21.4/20.9 m. The inner diameter is 18.5/18.8 m. The stator, the biggest of its kind, is 3.1/3 m in height. Bearing load is 5050/5500 tonnes. Average efficiency is over 94%, and reaches 96.5%.

Three Gorges Dam Francis turbine

The generators are manufactured by two joint ventures. One of them includes Alstom, ABB Group, Kvaerner, and the Chinese company Haerbin Motor. The other includes Voith, General Electric, Siemens (abbreviated as VGS), and the Chinese company Oriental Motor. The technology transfer agreement was signed together with the contract. Most of the generators are water-cooled. Some newer ones are air-cooled, which are simpler in design and manufacture and are easier to maintain.

Generator Installation Progress

The 14 north side main generators are in operation. The first (No. 2) started on July 10, 2003. The north side became completely operational September 7, 2005 with the implementation of generator No. 9. Full power (9,800 MW) was only reached on October 18, 2006 after the water level reached 156 m.

The 12 south side main generators are also in operation. No. 22 began operation on June 11, 2007 and No. 15 started up on October 30, 2008. The sixth (No. 17) began operation on December 18, 2007, raising capacity to 14.1 GW, finally surpassing Itaipu (14.0 GW), to become the world's largest hydro power plant by capacity.

The 6 underground main generators are also in operation as of May 23, 2012, when the last main generator, No. 27, finished its final test raising capacity to 22.5 GW. After 9 years of construction, installation and testing, the power plant is now fully operational.

Output Milestones

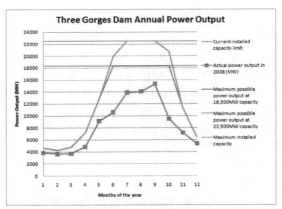

Three Gorges Dam annual power output

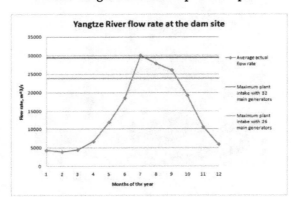

Yangtze River flow rate comparing to the dam intake capacity

By August 16, 2011, the plant had generated 500 TWh of electricity. In July 2008 it generated 10.3 TWh of electricity, its first month over 10 TWh. On June 30, 2009, after the river flow rate increased to over 24,000 m³, all 28 generators were switched on, producing only 16,100 MW because the head available during flood season is insufficient. During an August 2009 flood, the plant first reached its maximum output for a short period.

During the November to May dry season, power output is limited by the river's flow rate, as seen in the diagrams on the right. When there is enough flow, power output is limited by plant generating capacity. The maximum power-output curves were calculated based on the average flow rate at the dam site, assuming the water level is 175 m and the plant gross efficiency is 90.15%. The actual power output in 2008 was obtained based on the monthly electricity sent to the grid.

The Three Gorges Dam reached its design-maximum reservoir water level of 175 m (574 ft) for the first time on October 26, 2010, in which the intended annual power-generation capacity of 84.7 TWh was realized. In 2012, the dam's 32 generating units generated a record 98.1 TWh of electricity, which accounts for 14% of China's total hydro generation.

Annual Production of Electricity		
Year	Number of installed units	TWh
2003	6	8.607
2004	11	39.155
2005	14	49.090
2006	14	49.250
2007	21	61.600
2008	26	80.812
2009	26	79.470
2010	26	84.370
2011	29	78.290
2012	32	98.100
2013	32	83.270
2014	32	98.800
2015	32	87.000

Distribution

The State Grid Corporation and China Southern Power Grid paid a flat rate of ¥250 per MWh (US$35.7) until July 2, 2008. Since then, the price has varied by province, from ¥228.7–401.8 per MWh. Higher-paying customers receive priority, such as Shanghai. Nine provinces and two cities consume power from the dam.

Power distribution and transmission infrastructure cost about 34.387 billion Yuan. Construction was completed in December 2007, one year ahead of schedule.

Power is distributed over multiple 500 kilovolt (kV) transmission lines. Three Direct current (DC) lines to the East China Grid carry 7,200 MW: Three Gorges – Shanghai (3,000 MW), HVDC Three Gorges – Changzhou (3,000 MW), and HVDC Gezhouba – Shanghai (1,200 MW). The alternating current (AC) lines to the Central China Grid have a total capacity of 12,000 MW. The DC transmission line HVDC Three Gorges – Guangdong to the South China Grid has a capacity of 3,000 MW.

The dam was expected to provide 10% of China's power. However, electricity demand has increased more quickly than previously projected. Even fully operational, on average, it supports only about 1.7% of electricity demand in China in the year of 2011, when the Chinese electricity demand reached 4692.8 TWh.

Environmental Impact

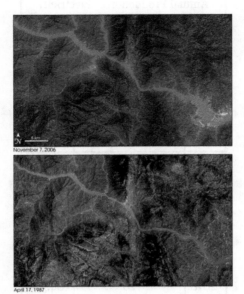

Satellite map showing areas flooded by the Three Gorges reservoir. Compare November 7, 2006 (above) with April 17, 1987 (below)

Flood mark on Yangtze river

Emissions

According to the National Development and Reform Commission of China, 366 grams of coal would produce 1 kWh of electricity during 2006. At full power, Three Gorges reduces coal consumption by 31 million tonnes per year, avoiding 100 million tonnes of greenhouse gas emissions, millions of tonnes of dust, one million tonnes of sulfur dioxide, 370,000 tonnes of nitric oxide, 10,000 tonnes of carbon monoxide, and a significant amount of mercury. Hydropower saves the energy needed to mine, wash, and transport the coal from northern China.

From 2003 to 2007, power production equaled that of 84 million tonnes of standard coal, reducing carbon dioxide by 190 million tonnes, sulfur dioxide by 2.29 million tonnes, and nitrogen oxides by 980,000 tonnes.

The dam increased the Yangtze's barge capacity sixfold, reducing carbon dioxide emission by 630,000 tonnes. From 2004 to 2007 a total of 198 million tonnes of goods passed through the ship locks. Compared to using trucking, barges reduced carbon dioxide emission by ten million tonnes and lowered costs by 25%.

Erosion and Sedimentation

Two hazards are uniquely identified with the dam. One is that sedimentation projections are not agreed upon, and the other is that the dam sits on a seismic fault. At current levels, 80% of the land in the area is experiencing erosion, depositing about 40 million tons of sediment into the Yangtze annually. Because the flow is slower above the dam, much of this sediment will now settle there instead of flowing downstream, and there will be less sediment downstream.

The absence of silt downstream has three effects:

- Some hydrologists expect downstream riverbanks to become more vulnerable to flooding.

- Shanghai, more than 1,600 km (990 mi) away, rests on a massive sedimentary plain. The "arriving silt—so long as it does arrive—strengthens the bed on which Shanghai is built... the less the tonnage of arriving sediment the more vulnerable is this biggest of Chinese cities to inundation..."

- Benthic sediment buildup causes biological damage and reduces aquatic biodiversity.

Earthquakes and Landslides

Erosion in the reservoir, induced by rising water, causes frequent major landslides that have led to noticeable disturbance in the reservoir surface, including two incidents in May 2009 when somewhere between 20,000 and 50,000 cubic metres (26,000 and 65,000 cu yd) of material plunged into the flooded Wuxia Gorge of the Wu River. Also, in the first four months of 2010, there were 97 significant landslides.

Waste Management

Zigui County seat source water protection area in Maoping Town, a few kilometers upstream of the dam

The dam catalyzed improved upstream wastewater treatment around Chongqing and its suburban areas. According to the Ministry of Environmental Protection, as of April 2007 more than 50 new plants could treat 1.84 million tonnes per day, 65% of the total need. About 32 landfills were added, which could handle 7,664.5 tonnes of solid waste every day. Over one billion tons of wastewater are released annually into the river, which was more likely to be swept away before the reservoir was created. This has left the water looking stagnant, polluted and murky.

Forest Cover

In 1997 the Three Gorges area had 10% forestation, down from 20% in the 1950s.

Research by the United Nations Food and Agriculture Organization research suggested that the Asia-Pacific region would, overall, gain about 6,000 km^2 (2,300 sq mi) of forest by 2008. That is quite a turnaround from the 13,000 km^2 (5,000 sq mi) net loss of forest each year in the 1990s. The main reason is China's huge reforestation effort. This accelerated after the 1998 Yangtze River floods convinced the government that it must restore tree cover, especially in the Yangtze's basin upstream of the Three Gorges Dam.

Wildlife

Concerns about the potential wildlife impact of the Dam predate the National People's Congress's approval in 1992. This region has long been known for its rich biodiversity. It is home to 6,388 species of plants, which belong to 238 families and 1508 genera. Of these plant species, 57 percent are endangered. These rare species are also used as ingredients in traditional Chinese medicines. Already, the percentage of forested area in the region surrounding the Three Gorges Dam has dropped from twenty percent in 1950 to less than ten percent as of 2002, negatively affecting all plant species in this locality. The region also provides habitats to hundreds of freshwater and terrestrial animal species. Freshwater fish are especially affected by dams due to changes in the water temperature and flow regime. Many other fish are hurt in the turbine blades of the hydro-electric plants as well. This is particularly detrimental to the ecosystem of the region because the Yangtze River basin is home to 361 different fish species and accounts for twenty-seven percent of all endangered freshwater fish species in China. Other aquatic species have been endangered by the dam, particularly the baiji, or Chinese river dolphin, now extinct. In fact, Government Chinese scholars even claim that the Three Gorges Dam directly caused the extinction of the baiji.

Of the 3,000 to 4,000 remaining critically endangered Siberian crane, a large number currently spend the winter in wetlands that will be destroyed by the Three Gorges Dam. The dam contributed to the functional extinction of the baiji Yangtze river dolphin. Though it was close to this level even at the start of construction, the dam further decreased its habitat and increased ship travel, which are among the factors causing what will be its ultimate demise. In addition, populations of the Yangtze sturgeon are guaranteed to be "negatively affected" by the dam.

Floods, Agriculture, Industry

An important function of the dam is to control flooding, which is a major problem for the seasonal river of the Yangtze. Millions of people live downstream of the dam, with many large, important cities like Wuhan, Nanjing, and Shanghai situated adjacent to the river. Plenty of farm land and China's most important industrial area are built beside the river.

The reservoir's flood storage capacity is 22 cubic kilometres (18,000,000 acre·ft). This capacity will reduce the frequency of major downstream flooding from once every ten years to once every 100 years. The dam is expected to minimize the effect of even a "super" flood. In 1954 the river flooded 193,000 km^2 (74,518 sq mi), killing 33,169 people and forcing 18,884,000 people to move. The flood covered Wuhan, a city of eight million people, for over three months, and the Jingguang

Railway was out of service for more than 100 days. The 1954 flood carried 50 cubic kilometres (12 cu mi) of water. The dam could only divert the water above Chenglingji, leaving 30 to 40 km³ (7.2 to 9.6 cu mi) to be diverted. Also the dam cannot protect against some of the large tributaries downstream, including the Xiang, Zishui, Yuanshui, Lishui, Hanshui, and the Gan.

In 1998 a flood in the same area caused billions of dollars in damage; 2,039 km² (787 sq mi) of farm land were flooded. The flood affected more than 2.3 million people, killing 1,526. In early August 2009, the largest flood in five years passed through the dam site. The dam limited the water flow to less than 40,000 cubic metres (52,000 cu yd) per second, raising the upstream water level from 145.13 metres on August 1, 2009, to 152.88 on August 8, 2009. 4.27 cubic kilometres of flood water were captured and the river flow was cut by as much as 15,000 cubic metres per second.

The dam discharges its reservoir during the dry season between December and March every year. This increases the flow rate of the river downstream, and provides fresh water for agricultural and industrial usage. It also improves shipping conditions. The water level upstream drops from 175 m to 145 m, preparing for the rainy season. The water also powers the Gezhouba Dam downstream.

Since the filling of the reservoir in 2003, the Three Gorges Dam has supplied an extra 11 cubic kilometres of fresh water to downstream cities and farms during the dry season.

During the 2010 South China floods, in July, inflows at the Three Gorges Dam reached a peak of 70,000 m³/s (2,500,000 cu ft/s), exceeding the peak during the 1998 Yangtze River Floods. The dam's reservoir rose nearly 3 m (9.8 ft) in 24 hours and reduced the outflow to 40,000 m³/s (1,400,000 cu ft/s) in discharges downstream, effectively alleviating serious impacts on the middle and lower river.

Navigating the Dam

Locks

Ship locks for river traffic to bypass the Three Gorges Dam, May 2004

The installation of ship locks is intended to increase river shipping from ten million to 100 million tonnes annually, as a result transportation costs will be cut between 30 and 37%. Shipping will become safer, since the gorges are notoriously dangerous to navigate. Ships with much deeper draft will be able to navigate 2,400 kilometres (1,500 mi) upstream from Shanghai all the way to Chongqing. It is expected that shipping to Chongqing will increase fivefold.

The other end of Three gorges dam lock, note the Bridge in the background

There are two series of ship locks installed near the dam (30°50′12″N 111°1′10″E30.83667°N 111.01944°E). Each of them is made up of five stages, with transit time at around four hours. Maximum vessel size is 10,000 tons. The locks are 280 m long, 35 m wide, and 5 m deep (918 × 114 × 16.4 ft). That is 30 m longer than those on the St Lawrence Seaway, but half as deep. Before the dam was constructed, the maximum freight capacity at the Three Gorges site was 18.0 million tonnes per year. From 2004 to 2007, a total of 198 million tonnes of freight passed through the locks. The freight capacity of the river increased six times and the cost of shipping was reduced by 25%. The total capacity of the ship locks is expected to reach 100 million tonnes per year.

These locks are staircase locks, whereby inner lock gate pairs serve as both the upper gate and lower gate. The gates are the vulnerable hinged type, which, if damaged, could temporarily render the entire flight unusable. As there are separate sets of locks for upstream and downstream traffic, this system is more water efficient than bi-directional staircase locks.

Ship Lift

The *shiplift*, a kind of elevator, can lift vessels of up to 3,000 tonnes, at a fraction of the time to transit the staircase locks.

In addition to the canal locks, there is a ship lift, a kind of elevator for vessels. The ship lift can lift ships of up to 3,000 tons. The vertical distance traveled is 113 metres, and the size of the ship lift's basin is 120×18×3.5 metres. The ship lift takes 30 to 40 minutes to transit, as opposed to the three to four hours for stepping through the locks. One complicating factor is that the water level can vary dramatically. The ship lift must work even if water levels vary by 12 meters (39 ft) on the lower side, and 30 metres on the upper side.

The ship lift's design uses a helical gear system, to climb or descend a toothed rack.

The ship lift was not yet complete when the rest of the project was officially opened on May 20, 2006. In November 2007 it was reported in the local media that construction of the ship lift started in October 2007.

In February 2012 *Xinhua* reported that the four towers that are to support the ship lift had almost been completed.

The report said the towers had reached 189 metres of the anticipated 195 metres, the towers would be completed by June 2012 and the entire shiplift in 2015.

As of May 2014, the ship lift was expected to be completed by July 2015. It was tested in December 2015 and announced complete in January 2016. Lahmeyer, the German firm that designed the ship lift, said it will take a vessel less than an hour to transit the lift. An article in Steel Construction says the actual time of the lift will be 21 minutes. It says that the expected dimensions of the 3,000 tonnes (3,000,000 kg) passenger vessels the ship lift's basin was designed to carry will be 84.5 metres (277 ft) X 17.2 metres (56 ft) X 2.65 metres (8.7 ft).

The trials of elevator finished in July 2016, the first cargo ship was lifted in July 15, the lift time comprised 8 minutes. *Shanghai Daily* reported that the first operational use of the lift was on September 18, 2016, when limited *"operational testing"* of the lift began.

Portage Railways

Plans also exist for the construction of short portage railways bypassing the dam area altogether. Two short rail lines, one on each side of the river, are to be constructed. The 88 kilometer long northern portage railway will run from the Taipingxi port facility on the northern side of the Yangtze, just upstream from the dam, via Yichang East Railway Station to the Baiyang Tianjiahe port facility in Baiyang Town, below Yichang. The 95 kilometer long southern portage railway will run from Maoping (upstream of the dam) via Yichang South Railway Station to Zhicheng (on the Jiaozuo–Liuzhou Railway).

In late 2012, preliminary work started along both future railway routes.

Relocation of Residents

As of June 2008, China relocated 1.24 million residents (ending with Gaoyang in Hubei Province) as 13 cities, 140 towns and 1350 villages either flooded or were partially flooded by the reservoir [A_2-M:CR3-1HP:S-15], about 1.5% of the province's 60.3 million and Chongqing Municipality's 31.44 million population. About 140,000 residents were relocated to other provinces.

Relocation was completed on July 22, 2008. Some 2007 reports claimed that Chongqing Municipality will encourage an additional four million people to move away from the dam to the main urban area of Chongqing by 2020. However, the municipal government explained that the relocation is due to urbanization, rather than the dam, and people involved included other areas of the municipality.

Allegedly, funds for relocating 13,000 farmers around Gaoyang disappeared after being sent to the local government, leaving residents without compensation.

Other Effects

Culture and Aesthetics

The 600 km (370 mi) long reservoir flooded some 1,300 archaeological sites and altered the appearance of the Three Gorges as the water level rose over 300 ft (91 m). Cultural and historical relics are being moved to higher ground as they are discovered, but the flooding inevitably covered undiscovered relics. Some sites could not be moved because of their location, size, or design. For example, the hanging coffins site high in the Shen Nong Gorge is part of the cliffs.

National Security

The United States Department of Defense reported that in Taiwan, "proponents of strikes against the mainland apparently hope that merely presenting credible threats to China's urban population or high-value targets, such as the Three Gorges Dam, will deter Chinese military coercion."

The notion that the military in Taiwan would seek to destroy the dam provoked an angry response from the mainland Chinese media. People's Liberation Army General Liu Yuan was quoted in the *China Youth Daily* saying that the People's Republic of China would be "seriously on guard against threats from Taiwan independence terrorists."

The three gorge dam is a steel-concrete gravity dam. The water is held back by the innate mass of the individual dam sections. As a result, damage to an individual section should not affect other parts of the dam. Due to the sheer size of the dam, it is expected to withstand tactical nuclear strikes.

Structural Integrity

Days after the first filling of the reservoir, around 80 hairline cracks were observed in the dam's structure. The submerged spillway gates of the dam might pose a risk of cavitation, similar to that which severely damaged the poorly designed and cavitating spillways of the Glen Canyon Dam in the US state of Arizona, which was unable to properly withstand the Colorado river floods of 1983. However 163,000 concrete units of the Three Gorges dam all passed quality testing and the deformation was within design limits. An experts group gave the project overall a good quality rating.

Upstream Dams

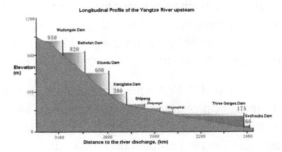

Longitudinal profile of upstream Yangtze River

In order to maximize the utility of the Three Gorges Dam and cut down on sedimentation from the

Jinsha River, the upper course of the Yangtze River, authorities plan to build a series of dams on the Jinsha, including Wudongde Dam, Baihetan Dam, along with the now completed Xiluodu and Xiangjiaba dams. The total capacity of those four dams is 38,500 MW, almost double the capacity of the Three Gorges. Baihetan is preparing for construction and Wudongde is seeking government approval. Another eight dams are in the midstream of the Jinsha and eight more upstream of it.

W. A. C. Bennett Dam

The W. A. C. Bennett Dam is a large hydroelectric dam on the Peace River in northern British Columbia, Canada. At 183 m (660 ft) high, it is one of the world's highest earth fill dams. Construction of the dam began in 1961 and culminated in 1968. At the dam, the Finlay, the Parsnip and the Peace Rivers feed into Williston Lake, also referred to as Williston Reservoir. It is the third largest artificial lake in North America (after the Smallwood Reservoir and Manicouagan) as well as the largest body of fresh water in British Columbia. Williston Lake runs 250 kilometres north-south and 150 kilometres east-west.

The construction of the dam cost $750 million, making it the largest project of its kind in the province of BC. The dam was named after the premier because his vision played a major role in the project initiation, development and realization; the reservoir was named after the premier's trusted cabinet colleague Ray Williston. The Gordon M. Shrum Generation Station at the W.A.C. Bennett Dam has the capacity to generate more than 13 billion kWh annually. At the time of its construction the powerhouse was the largest of its kind worldwide. In addition to the benefits related to the energy generated, the construction of the dam and the reservoir also provided economic opportunities for the province of British Columbia, for the newly founded provincially owned electric utility BC Hydro, and for the large number of workers. These workers were involved in the planning, construction, operation, and maintenance of the project. Considerable costs were involved in the government funded project, the clearing of the area for the reservoir, called the Trench, alone cost $5 million.

The building of the dam and the reservoir were not without controversy. One controversy was caused by the significant negative environmental effects the project had on the immediate environment. In the process of creating Williston Lake, 350,000 acres of forested land was flooded. This caused the loss of plant and wildlife biodiversity as well as the loss of minerals and timber rights.

A second controversy related to the fact that the land had been inhabited prior to its flooding, therefore the flooding resulted in the displacement of the residents located in the Trench. Among them were members of the Tsay Keh Dene First Nation, then known as Ingenika. The displacement had negative social impacts on the inhabitants as the loss of the land that had previously supported them meant loss of autonomy and resulted in isolation, alienation, and "social disorganization". A BC Hydro consultant admitted in 1977 that the 'isolation imposed by the reservoir had severe impacts on Ingenika society and culture".

From 2009 to 2012, units 6 to 8 were refurbished to increase the plant capacity by 90 megawatts. The last unit of them, Unit 7, was put into service on November 29, 2012.

Units 1 to 5 have undergone complete refurbishment and upgrades in the last three years, increasing the reliability and capacity of the first units installed at GM Shrum. The upgrades increase the generating capacity of each unit by as much as 17%.

History

W.A.C. Bennett and High Modernism

W.A.C. Bennett was the Premier of British Columbia from 1952 to 1972. Bennett was committed to the large-scale, state directed development of British Columbia and promoted the continued development of natural resources. Large hydroelectric projects, such as the W.A.C. Bennett Dam, were part of the resource development for which Bennett was advocating. In his opinion, harnessing nature would make British Columbia wealthy and support the emergence of an industrial economy as well as a society that was, "connected, institutionally anchored, urban, wealthy, and domestic.".

Bennett's convictions, and therefore the policies of his government, concerning hydroelectric development have been regarded as a manifestation of the ideology high modernity, also known as high modernism. Along with the benefits that high modernist development could bring, there were also consequences. High modernism, along with the administrative ordering of nature and society, authoritarian state, and a "prostrate civil society which would be unable to resist high modernist plans", can be a recipe for disaster. It is debatable whether or not all of these elements were present in British Columbia at the time, but regardless, the development of the Peace River led to environmental changes that caused a minority of people to live in isolation, dependence, alienation, and illness. On the other hand, the hydroelectric projects realized by Bennett's Two Rivers policy created a large supply of less expensive energy in British Columbia, which provided industrial growth and therefore employment.

Two Rivers Policy

W.A.C. Bennett's Two Rivers policy aimed to develop the hydroelectric potential of both the Peace and Columbia Rivers simultaneously. The policy stemmed from Bennett's desire to wrest control of resources away from the federal government in regards to power development in the province. Bennett and the American company Kaiser Aluminum and Chemical Corporation of the United States had agreed that in return for a fifty-year water license, the Kaiser Corporation would construct a large dam on the upper Columbia River. Not only would they pay for the construction, they would also return 20% of the electricity generated to British Columbia and pay the provincial taxes as well as water license fees. Much to Bennett's dismay, the federal government of Canada dissolved the deal by asserting its right of control over international waterways and took over negotiations with the United States. It would appear that British Columbia was not to be allowed to sell electricity to America for provincial profit.

In response, Bennett turned his attention towards developing the Peace River's hydroelectric potential at the previously identified site of Portage Mountain by constructing a massive storage dam that would later be named the W.A.C. Bennett dam. Bennett hoped that the economic independence British Columbia would gain by developing the Peace River would provide the leverage necessary for the federal government of Canada to allow British Columbia to sell electricity that could be created by damming the Columbia River to the Americans. This 'Two Rivers' policy experienced opposition from people who thought that if the Columbia were developed, the electricity generated should be for Canada's sole use as opposed to America's.

The Two Rivers policy led to the development of the WAC Bennett Dam on the Peace River and the Keenleyside Dam and Mica Dam on the Columbia River together with Duncan Dam at the top of Kootenay Lake. In 1964 the policy was formally realized with ratification of the Columbia River Treaty by the Government of Canada and the United States of America. Because of his Two Rivers policy, Bennett was successful in pressuring the federal government of Canada to allow British Columbia to 'sell electricity' to the Americans for a thirty-year period for the lump sum of US$275 million. The nationalization of BC Electric (1961), which was rolled over into the BC Power Commission to form BC Hydro, can also be seen as a part of this strategy.

Site Selection

Ray Williston, the minister of land and forests for the provincial government at the time, proposed turning sections of the Peace and Columbia River basins known as the "rocky mountain trench" into a source of power generation. The 'Power Trench', as it was known, would provide not only electricity, but give the ability to control water flow for flood prevention and agricultural purposes in the U.S. and Canada. In 1957 twelve locations along the Peace River were identified by the Wenner-Gren British Columbia Development Company as potential sites to build a dam. One of the sites, located 22 kilometres from Hudson's Hope, was judged to be the best location due to its geography. Gordon Shrum, a physics professor at the University of British Columbia, was chosen to conduct a study on the cost effectiveness of developing dams on the Peace and Columbia rivers. The study led to the conclusion that it would be cheaper to build on the Peace River, but only if a public company was used due to lower interest rates available to crown corporations.

Construction

When plans for construction were given the green light on the W.A.C. Bennett Dam (known as the Portage Mountain Dam during construction), clearing the soon to be reservoir was the first step in the process. It was a massive undertaking which was completed on the shoestring budget of five million dollars by the Forest Service Branch. The initial stages of construction required building a coffer dam, preparing the foundations and injecting grout into the ground to create a watertight seal, building a drainage system, and excavating to create a solid base for building. Over the course of construction 55 million cubic yards of rock and dirt were taken from the nearby glacial moraine by conveyor belt to create the dam relying primarily on gravity to hold it together. Upon completion, the W.A.C. Bennett Dam became one of the biggest earth filled dams in the world stretching 183 metres tall, 800 metres wide, by two kilometres long. When finished,the dam incorporated one of the largest hydroelectric generating stations. Components are located as far as 150 metres below ground and includes 10 generating units located deep underground in the powerhouse. The Main powerhouse structure is named the G.M. Shrum generating station. It was designed to resemble a giant transformer to reflect its function and modern design of the 1960s.

The project was widely seen as a success, especially considering its remote location far from civilization. The construction project was managed by Gordon Shrum, the appointed head of the newly created BC Hydro crown corporation. The provincial government had specifically created BC Hydro as a way of financing the project through lower interest rates available to crown corporations and to control the development of provincial energy resources. When Shrum took over the project in 1961 it was already a year behind schedule meeting the 1968 deadline to generate power. Through a 'hands on', 'cost conscious' and a 'design as you go' strategy, the project was officially completed in the fall of 1967 with the first generators going online in 1968. The project was completed on time and on budget, however additional construction would continue through the 1970s with final completion in 1980 when the last generator was installed.

The construction of the W.A.C. Bennett Dam involved over twenty unions that were bound by ten-year contracts guaranteeing BC Hydro no lockout (industry) or strike action. This contract allowed the project to be built without labour delays. The men involved on the project were international, coming from around North America, Europe, and as far as Japan. The workers onsite lived in temporary camps built around the Portage Mountain site with more workers in the summer and less in the winter. Much of the construction occurred inside the dam which was claustrophobic, full of exhaust fumes, and occasionally subject to cave-ins. In total, 16 men have lost their lives working on the dam.

During the construction process, the portage mountain lookout was one of the first buildings built so that tourists could view the progress on the dam.

Economic Investment and Opportunity

Province of British Columbia

In the 1950s, as well as the decades before and after, the economy of British Columbia had largely been based on the extraction of natural resources and had therefore been susceptible to fluctuations in the world's demand for the respective resources. Despite the potentially unreliable economy resulting from this susceptibility, British Columbia was considered to be one of the most sought out Canadian provinces to live in. This was due to the fact that British Columbia had the country's highest real per capita income which resulted in high standards of living for its residents. It was not until W.A.C. Bennett's premiership and vision for his province though that British Columbia saw the realization of its hydroelectric energy potential. Bennett believed that any natural resource that was not used was wasted and pushed for the development of ways to harness the enormous unrealized hydroelectric energy power potential of the Peace River. Today, the W.A.C. Bennett and Peace Canyon facilities produces about 35% of British Columbia's total electricity.

BC Hydro

The British Columbia Hydro and Power Authority Act, introduced by Premier W.A.C. Bennett in March 1962, laid out the plan in which he would pursue his Two Rivers Policy. BC Electric had refused to commit to buying the power that would be harnessed from the Peace River development as cheaper power was available elsewhere. Hence, the BC Hydro and Power Authority Act merged BC Electric with another crown corporation, the BC Power Commission, into a newly formed BC Hydro which was co-chaired by Gordon Shrum of BC Electric and Hugh Keenleyside of BC Power Commission. BC Hydro became responsible for the building of the dam, powerhouse and associated infrastructure.

Local Community and Workers

The building of the dam and the powerhouse and the creation of Williston Lake provided economic opportunities to the high number of workers who found employment with BC Hydro or one of the subcontractors; these workers included members of the local first nations, non-native residents, non-residents, and immigrants. One of the subcontractors was the Forest Service Branch of the Department of Lands and Forests to whom BC Hydro paid $5 million to clear the area that would become Willison Lake - an area that was covered in timber to 80%. At the peak of project, 3500 workers were employed. Many of them had moved to the area for the job and settled down, at least temporarily, in close proximity to the dam project. Hudson's Hope, a frontier town, was one of the communities in which many of the non-resident workers found a home; during the project, the population of Hudson's Hope rose from 800 to over 5000 in 1968 and dropped to less than 1500 by the early 1980s. In addition, about 2000 workers lived at a camp in close proximity to the dam.

Social Impacts

Aboriginal Communities

For a minority of people, many of whom were Aboriginal, environmental changes caused by the damming of the Peace River meant dependence, isolation, alienation, and illness. When it became clear that the environmental impacts of the dam would render land unlivable to local Aboriginal groups who were dependent on the hereditary sites, the British Columbia government offered a settlement. For 1.7 million dollars the British Columbia government purchased fourteen thousand acres of land, including timber and mineral rights, and bought out approximately one hundred and fifty individuals and families securing the rights to the land. Of those one hundred and fifty, roughly one third were members of the Tsay Keh Dene First Nation. Outside of relocation, Aboriginal hunting and fishing grounds around the Fort Grahame and Finlay Forks areas were severely impacted by ecological change. Many species of fish as well as mountain caribou and muskrats were no longer available for Aboriginal consumption or traditional use. These changes to First Nations independence through fur trade and the relocation of many Aboriginals to new reserves caused an influx in demand for government assistance through welfare. Between 1965 and 1970, social assistance provided by the Provincial government to Aboriginal groups in the areas surrounding the Bennett Dam increased by 300 percent.

As recently as October 2008, the Kwadacha First Nation, another Aboriginal group residing in the Fort Ware area located at the north end of the Finlay Reach of Lake Williston, reached a settlement with the British Columbia government and BC Hydro over damages suffered during construction and operation of the dam and Williston Lake. The settlement included a $15 million lump-sum payment and annual payments of $1.6 million adjusted for inflation.

Local Residents

When the government, controlled by the Social Credit Party of British Columbia, dammed the Peace River to generate hydroelectricity it had already set into motion a series of social changes in the surrounding communities. These social changes had positive effects for workers who flocked to the area to secure jobs and economic opportunities unavailable elsewhere. It also had negative effects for residents who lived in the surrounding areas prior to the dam's construction. British Columbia Premier WAC Bennett saw growing communities when he envisioned the damming of the Peace River in 1952. In 1964, his vision would be validated as a result of the "instant town" of Mackenzie, where thousands of individuals would find employment in the forest industry which was a direct result of clearing land for the construction of the dam. For residents who had lived in the surrounding areas prior to the dams planned construction, development caused many to be pushed off homesteads for small monetary settlements. One resident who owned a thousand acres of land, much of which was used for farming, was offered only twenty-eight thousand dollars by BC Hydro to secure the property. However, for local residents of Anglo ethnicity, full-time waged work was more easily accessible due to the employment opportunities produced directly and indirectly by the damming project.

Environmental Impacts

Downstream

The W.A.C. Bennett Dam held tremendous economic potential, but for its surrounding environment the experience was not so positive. Since its construction a number of environmental changes have taken place. The dam has been responsible for drastic fluctuations in the water levels of the upstream and downstream portions of the Peace River, creating modifications to both the plants and animals in the region. In addition, it has also been blamed for creating changes in the landscapes of the Athabasca Lake and Peace River, known as the Athabasca Delta . This part of the river faced significant water loss. While the area of the Peace River immediately upstream of the dam was experiencing flooding, which gave rise to Williston Lake, downstream the Peace-Athabasca Delta was drying up . For this reason, the delta experienced several changes in the water level, affecting both the landscape of the delta and its aquatic life.

Following the completion of the Williston Lake in 1971, water coverage was reduced to 38 percent and the amount of wetlands and wet marches declined to 47 percent . Floods that occurred every two or three years came to a halt, no longer able to revitalize the biodiversity alongside the delta. A reduction in the amount of discharge resulted in the accumulation of toxins and sediments downstream, decreasing the quality of the water. Fish also experienced changes as a result of the low water levels: fewer channels were accessible for walleye to reach spawning grounds and for juvenile fish to reach nursery areas therefore jeopardizing their ability to reproduce. Dinosaur Lake was created directly downstream of the W.A.C. Bennett Dam when the Peace Canyon Dam was completed. The Peace Canyon Dam was built to maximize the generation of hydro-electricity that the W.A.C. Bennett Dam couldn't capture. Today, it is a popular destination for camping in British Columbia.

Upstream and Williston Lake

The area upstream of the dam experienced a number of environmental changes as a result of the flooding of the land. The creation of the lake flooded a vast area of forested land, drowning a significant amount of wildlife and creating drastic changes to the landscape and climate. It created a reservoir that measured 250 kilometres from north to south and another 150 kilometers from east to west. Farmers had asked for compensation from BC Hydro because the changes created in the weather compromised their ability to grow crops. Because the water was no longer flowing, rather standing still following the creation of the dam there was an increase in humidity in the area. This caused cooler temperatures and an increase in fog.

Not only did changes occur in the atmosphere, they also occurred in the water. The creation of the reservoir compromised the livelihood of aquatic life, which before the dam lived peacefully in the river . Rivers and lakes support different species therefore some fish were able to thrive in the lake but others could not be supported by its different environment. Mountain whitefish, rainbow trout and Arctic grayling were primarily the species that faced decline. A number of species were known to have thrived and it is estimated that there are more fish in the basin today than before the reservoir, but scientists indicate they are not entirely healthy. High levels of mercury have been measured in the lake, as a result of decaying matter from the plants and trees that drowned. Mercury accumulates in the lake, is ingested by tiny organisms and eventually makes its way up the food chain. In the year 2000 British Columbia issued a Fish Consumption Advisory for bull trout and dolly varden warning people about the high content of mercury in these fish .

Visitor's Centre

The W.A.C. Bennett Dam Visitor Centre is located near the dam, overlooking Williston Lake Reservoir. The centre features exhibits on the dam, hydroelectricity, and the area's natural and cultural history.

Itaipu Dam

The Itaipu Dam is a hydroelectric dam on the Paraná River located on the border between Brazil and Paraguay. The name "Itaipu" was taken from an isle that existed near the construction site. In the Guarani language, *Itaipu* means "the sounding stone". Completed in 1984, it is a binational undertaking run by Brazil and Paraguay at the border between the two countries, 15 km (9.3 mi) north of the Friendship Bridge. The project ranges from Foz do Iguaçu, in Brazil, and Ciudad del Este in Paraguay, in the south to Guaíra and Salto del Guairá in the north. The installed generation capacity of the plant is 14 GW, with 20 generating units providing 700 MW each with a hydraulic design head of 118 metres (387 ft). In 2013 the plant generated a record 98.6 TWh, supplying approximately 75% of the electricity consumed by Paraguay and 17% of that consumed by Brazil.

Of the twenty generator units currently installed, ten generate at 50 Hz for Paraguay and ten generate at 60 Hz for Brazil. Since the output capacity of the Paraguayan generators far exceeds the load in Paraguay, most of their production is exported directly to the Brazilian side, from where two 600 kV HVDC lines, each approximately 800 kilometres (500 mi) long, carry the majority of the energy to the São Paulo/Rio de Janeiro region where the terminal equipment converts the power to 60 Hz.

Entrance Itaipu dam

History

Negotiations between Brazil and Paraguay

The concept behind the Itaipu Power Plant was the result of serious negotiations between the two countries during the 1960s. The "Ata do Iguaçu" (Iguaçu Act) was signed on July 22, 1966, by the Brazilian and Paraguayan Ministers of Foreign Affairs, Juracy Magalhães and Raúl Sapena Pastor, respectively. This was a joint declaration of the mutual interest in studying the exploitation of the hydro resources that the two countries shared in the section of the Paraná River starting from, and including, the *Salto de Sete Quedas*, to the *Iguaçu River* watershed. The Treaty that gave origin to the power plant was signed in 1973.

The terms of the treaty, which expires in 2023, have been the subject of widespread discontent in Paraguay. The government of President Lugo vowed to renegotiate the terms of the treaty with Brazil, which long remained hostile to any renegotiation.

In 2009, Brazil agreed to a fairer payment of electricity to Paraguay and also allowed Paraguay to sell excess power directly to Brazilian companies instead of solely through the Brazilian electricity monopoly.

Construction Starts

In 1970, the consortium formed by the companies IECO (from the United States), and ELC Electroconsult S.p.A. (from Italy) won the international competition for the realization of the viability studies and for the elaboration of the construction project. Design studies began in February 1971. On April 26, 1973, Brazil and Paraguay signed the Itaipu Treaty, the legal instrument for the hydroelectric exploitation of the Paraná River by the two countries. On May 17, 1974, the Itaipu Binacional entity was created to administer the plant's construction. The construction began in January of the following year. Brazil's (and Latin America's) first electric car was introduced in late 1974; it received the name *Itaipu* in honor of the project.

Paraná River Rerouted

On October 14, 1978, the Paraná River had its route changed, which allowed a section of the riverbed to dry so the dam could be built there.

Agreement by Brazil, Paraguay, and Argentina

An important diplomatic settlement was reached with the signing of the *Acordo Tripartite* by Brazil, Paraguay and Argentina, on October 19, 1979. This agreement established the allowed river levels and how much they could change as a result of the various hydroelectrical undertakings in the watershed that was shared by the three countries. At that time, the three countries were ruled by military dictatorships. Argentina was concerned that, in the event of a conflict, Brazil could open the floodgates, raising the water level in the Río de la Plata and consequently flood the capital city of Buenos Aires.

Formation of the Lake

The reservoir began its formation on October 13, 1982, when the dam works were completed and the side canal's gates were closed. Throughout this period, heavy rains and flooding accelerated the filling of the reservoir as the water rose 100 meters (330 feet) and reached the gates of the spillway at 10:00 on October 27.

Start of Operations

On May 5, 1984, the first generation unit started running in Itaipu. The first 18 units were installed at the rate of two to three a year; the last two of these started running in the year 1991.

Capacity Expansion in 2007

The dam undergoes expansion work.

The last two of the 20 electric generation units started operations in September 2006 and in March 2007, thus raising the installed capacity to 14 GW and completing the power plant. This increase in capacity allows 18 generation units to run permanently while two are shut down for maintenance. Due to a clause in the treaty signed between Brazil, Paraguay and Argentina, the maximum number of generating units allowed to operate simultaneously cannot exceed 18.

The rated nominal power of each generating unit (turbine and generator) is 700 MW. However, because the head (difference between reservoir level and the river level at the foot of the dam) that actually occurs is higher than the designed head (118 m), the power available exceeds 750 MW half of the time for each generator. Each turbine generates around 700 MW; by comparison, all the water from the Iguaçu Falls would have the capacity to feed only two generators.

November 2009 Power Failure

On November 10, 2009, transmission from the plant was completely disrupted, possibly due to a storm damaging up to three high-voltage transmission lines. Itaipu itself was not damaged. This caused massive power outages in Brazil and Paraguay, blacking out the entire country of Paraguay for 15 minutes, and plunging Rio de Janeiro and São Paulo into darkness for more than 2 hours. 50 million people were reportedly affected. The blackout hit at 22:13 local time. It affected the southeast of Brazil most severely, leaving São Paulo, Rio de Janeiro and Espírito Santo completely without electricity. Blackouts also swept through the interior of Rio Grande do Sul, Santa Catarina, Mato Grosso do Sul, Mato Grosso, the interior of Bahia and parts of Pernambuco, energy officials said. By 00:30 power had been restored to most areas.

Wonder of the Modern World

In 1994, the American Society of Civil Engineers elected the Itaipu Dam as one of the seven modern Wonders of the World. In 1995, the American magazine *Popular Mechanics* published the results.

Panoramic view of the Itaipu Dam, with the spillways (closed at the time of the photo) on the left

This Diagram shows in detail the heights:

325 metres (1,066 ft), entire dam including the 100 metres (330 ft) high Power Line 4 Pylons atop the Barrage
260 metres (850 ft), dam + the foundation inside water until the river floor
247 metres (810 ft), 196 metres (643 ft) high of roof reinforcement concrete dam + Cranes atop the Barrage
225 metres (738 ft), Elevation End Main Concrete Barrage

196 metres (643 ft), The official Roof given from Itaipú Binacional Webpage

Social and Environmental Impacts

When construction of the dam began, approximately 10,000 families living beside the Paraná River were displaced, because of construction.

The world's largest waterfall by volume, the Guaíra Falls, was drowned by the newly formed Itaipu reservoir. The Brazilian government liquidated the Guaíra Falls National Park, and dynamited the submerged rock face where the falls had been, facilitating safer navigation, thus eliminating the possibility of restoring the falls in the future. A few months before the reservoir was filled, 80 people died when an overcrowded bridge overlooking the falls collapsed, as tourists sought a last glimpse of the falls.

The American composer Philip Glass has also written a symphonic cantata named *Itaipu*, in honour of the structure.

The Santa Maria Ecological Corridor now connects the Iguaçu National Park with the protected margins of Lake Itaipu, and via these margins with the Ilha Grande National Park.

Statistics

Central Control Room (CCR)

Itaipu penstocks

Construction

- The course of the seventh biggest river in the world was shifted, as were 50 million tons of earth and rock.

- The amount of concrete used to build the Itaipu Power Plant would be enough to build 210 football stadiums the size of the Estádio do Maracanã.

- The iron and steel used would allow for the construction of 380 Eiffel Towers.

- The volume of excavation of earth and rock in Itaipu is 8.5 times greater than that of the Channel Tunnel and the volume of concrete is 15 times greater.

- Around forty thousand people worked in the construction.

- Itaipu is one of the most expensive objects ever built.

Generating Station and Dam

- The total length of the dam is 7,235 metres (23,737 ft). The crest elevation is 225 metres (738 ft). Itaipu is actually four dams joined together — from the far left, an earth fill dam, a rock fill dam, a concrete buttress main dam, and a concrete wing dam to the right.

- The spillway has a length of 483 metres (1,585 ft).

- The maximum flow of Itaipu's fourteen segmented spillways is 62.2 thousand cubic metres per second (2.20×10^6 cu ft/s), into three skislope formed canals. It is equivalent to 40 times the average flow of the nearby natural Iguaçu Falls.

- The flow of two generators (700 cubic metres per second (25,000 cu ft/s) each) is roughly equivalent to the average flow of the Iguaçu Falls (1,500 cubic metres per second (53,000 cu ft/s)).

- If Brazil were to use Thermal Power Generation to produce the electric power of Itaipu, 434,000 barrels (69,000 m³) of petroleum would have to be burned every day.

- The dam is 196 metres (643 ft) high, equivalent to a 65-story building.

- Though it is the seventh largest reservoir in size in Brazil, the Itaipu's reservoir has the best relation between electricity production and flooded area. For the 14,000 MW installed power, 1,350 square kilometres (520 sq mi) were flooded. The reservoirs for the hydroelectric power plants of Sobradinho Dam, Tucuruí Dam, Porto Primavera Dam, Balbina Dam, Serra da Mesa Dam and Furnas Dam are all larger than the one for Itaipu, but have a smaller installed generating capacity. The one with the largest hydroelectric production, Tucuruí, has an installed capacity of 8,000 MW, while flooding 2,430 km² (938 sq mi) of land.

- Electricity is 55% cheaper when made by the Itaipu Dam than the other types of power plants in the area.

Tarbela Dam

Tarbela Dam is an earth fill dam located on the Indus River in Pakistan. It is the largest earth-filled dam in the world and fifth-largest by structural volume. It is named after the town Tarbela, in Khyber Pakhtunkhwa, about 50 kilometres (31 mi) northwest of Islamabad.

The dam is 485 feet (148 m) high above the riverbed. The dam forms the Tarbela Reservoir, with a surface area of approximately 250 square kilometres (97 sq mi). The dam was completed in 1976 and was designed to store water from the Indus River for irrigation, flood control, and the generation of hydroelectric power.

The primary use of the dam is for electricity generation, the installed capacity of the 3,478 MW Tarbela hydroelectric power stations will increase to 6,298MW after completion of the ongoing fourth extension and the planned fifth extension financed by Asian Infrastructure Investment Bank and the World Bank.

Project Description

The project is located at a narrow spot in the Indus River valley, at Tarbela between Haripur District and Swabi District, approximately 60 kilometers northwest of Islamabad.

The main dam wall, built of earth and rock fill, stretches 2,743 metres (8,999 ft) from the island to river right, standing 148 metres (486 ft) high. A pair of concrete auxiliary dams spans the river from the island to river left. The dam's two spillways are located on the auxiliary dams rather than the main dam. The main spillway has a discharge capacity of 18,406 cubic metres per second

(650,000 cu ft/s) and the auxiliary spillway, 24,070 cubic metres per second (850,000 cu ft/s). Annually, over 70% of water discharged at Tarbela passes over the spillways, and is not used for hydropower generation.

Tarbela Dam

Five large tunnels were constructed as part of Tarbela Dam's outlet works. Hydroelectricity is generated from turbines in tunnel 1 through 3, while tunnels 4 and 5 were designed for irrigation use, although both tunnels are to be converted to hydropower tunnels to increase Tarbela's electricity generating capacity. These tunnels were originally used to divert the Indus River while the dam was being constructed.

MA hydroelectric power plant on the right side of the main dam houses 14 generators fed with water from outlet tunnels 1, 2, and 3. There are four 175 MW generators on tunnel 1, six 175 MW generators on tunnel 2, and four 432 MW generators on tunnel 3, for a total generating capacity of 3,478 MW.

Tarbela Reservoir is 80.5 kilometres (50.0 mi) long, with a surface area of 250 square kilometres (97 sq mi). The reservoir initially stored 11,600,000 acre feet (14.3 km³) of water, with a live storage of 9,700,000 acre feet (12.0 km³), though this figure has been reduced over the subsequent 35 years of operation to 6,800,000 acre feet (8.4 km³) due to silting.

The catchment area upriver of the Tarbela Dam is spread over 168,000 square kilometres (65,000 sq mi) of land largely supplemented by snow and glacier melt from the southern slopes of the Himalayas. There are two main Indus River tributaries upstream of the Tarbela Dam. These are the Shyok River, joining near Skardu, and the Siran River near Tarbela.

Background

Tarbela dam was constructed as part of the Indus Basin Project after signing of the 1960 Indus Waters Treaty between India and Pakistan. The purpose was to compensate for the loss of water supplies of the eastern rivers (Ravi, Sutlej and Beas) that were designated for exclusive use by India per terms of the Treaty. The primary objective of the dam was to supply water for irrigation by storing flows during the monsoon period and subsequently releasing stored water during the low flow period in winter.

By the mid 1970s, power generation capacity was added in three subsequent hydro-electrical project extensions which were completed in 1992, installing a total of 3,478 MW generating capacity on respectively Tunnel 1 (four turbines), Tunnel 2 (six turbines) and Tunnel 3 (four turbines).

Construction

Construction of Tarbela Dam was carried out in three stages to meet the diversion requirements of the river. Construction was undertaken by the Italian firm Salini Impregilo.

Stage 1

In the first stage, the Indus river was allowed to flow in its natural channel, while construction works commenced on the right bank where a 1500 feet (457 meters) long and 694 feet (212 meters) wide diversion channel was being excavated along with a 105 feet (32 meters) high buttress dam that was also being construction. Stage 1 construction lasted approximately 2.5 years.

Stage 2

The main embankment dam and the upstream blanket were constructed across the main valley of the river Indus as part of the second stage of construction. During this time, water from the Indus river remained diverted through the diversion channel. By the end of construction works in stage 2, tunnels had been built for diversion purposes. Stage 2 construction took 3 years to complete.

Stage 3

Under the third stage of construction, works were carried out on the closure of diversion channel and construction of the dam in that portion while the river was made to flow through diversion tunnels. The remaining portion of upstream blanket and the main dam at higher levels was also completed as part of stage 3 works, which were concluded in 1976.

Re-settlement of People Affected by Tarbela Dam

It has a reservoir area of about 260 square kilometers and about 82,000 acres (33,000 ha) of land was acquired for its construction. The large reservoir of the dam submerged 135 villages, which resulted in displacement of a population of about 96,000 people, many of whom were relocated to a series of townships surrounding the Tarbela reservoir or in adjacent higher valleys.

For the land and built-up property acquired under the Land Acquisition Act of 1984, a cash compensation of Rs 469.65 million was paid to those affected. In the absence of a national policy, re-settlement concerns of the affectees of Tarbela Dam were addressed on an ad hoc basis.

Many affectees of the Tarbela Dam have still not been settled and have not been given land as compensation by the government of Pakistan, in accordance with its contractual obligations with the World Bank. However, some of the affectees have been settled at Khalabat Township and other places across Pakistan.

Lifespan

Because the source of the Indus River is glacial meltwater from the Himalayas, the river carries huge amounts of sediment, with an annual suspended sediment load of 200 million tons.

Live storage capacity of Terbela reservoir had declined more than 33.5 per cent to 6.434 million acre feet (MAF) against its original capacity of 9.679 MAF because of sedimentation over the past 38 years. The useful life of the dam and reservoir was estimated to be approximately fifty years. However, sedimentation has been much lower than predicted, and it is now estimated that the useful lifespan of the dam will be 85 years, to about 2060.

Pakistan also plans to construct several large dams upstream of Tarbela, including the Diamer-Bhasha Dam. Upon completion of the Diamer-Bhasha dam, sediment loads into Tarbela will be decreased by 69%.

Project Benefits

In addition to fulfilling the primary purpose of the dam, i.e. supplying water for irrigation, Tarbela Power Station has generated 341.139 billion kWh of cheap hydro-electric energy since commissioning. A record annual generation of 16.463 billion kWh was recorded during 1998–99. Annual generation during 2007–08 was 14.959 billion kWh while the station shared peak load of 3702 MW during the year, which was 23.057% of total WAPDA system peak.

Tarbela-IV Extension Project

In September 2013, Pakistan's Water and Power Development Authority signed a Rs. 26.053 billion contract with Chinese firm Sinohydro and Germany's Voith Hydro for executing civil works on the 1,410 MW *Tarbela-IV Extension Project*. Construction commenced in February 2014, and is expected to be completed by June 2017. Upon completion, the project will augment the Early Harvest Project of the China–Pakistan Economic Corridor.

The project is being constructed at tunnel 4 of the dam which was originally intended for irrigation, but was later taken up for power generation as country's electricity demand increased. As much of the infrastructure for the project already exists, conversion of the irrigation tunnel into a hydroelectric tunnel was regarded as a cost-effective way to ease Pakistan's energy shortfall.

Annual benefits of the project were estimated at about Rs. 30.7 billion. On an annual basis, over 70% of water passing through Tarbela is discharged over spillways, while only a portion of the remaining 30% is used for hydropower generation.

As of February 2016, the Government of Pakistan states that 49% of works have been completed on the 4th phase extension project.

Financing

The project's cost was initially estimated to be $928 million, but the cost was revised downwards to $651 million. The World Bank had agreed to provide a $840 million loan for the project in June 2013. The loan had two components: The first component is a $400 million International Development Association loan, which will be lent as a concessional loan at low interest rates. The second portion consists of a $440 million from the World Bank's International Bank for Reconstruction and Development. Pakistan's Water and Power Development Authority was to provide the remaining $74 million required for construction, before the project's cost was downwardly revised by $277 million. Interest costs for the loans are estimated to cost $83.5 million.

Because of revised lower costs to $651 million from $928 million, the World Bank permitted Pakistani officials to expedite completion of the project by 8 months at a cost of an additional $51 million. Pakistani officials were also permitted to divert $126 million towards the Tarbela-V Extension Project.

Tarbela-V Extension Project

The Tarbela Dam was built with five original tunnels, with the first three dedicated to hydropower generation, and the remaining two slated for irrigation use; the fourth phase extension project utilizes the first of the two irrigation tunnels, while the 5th phase extension will use the second irrigation tunnel. Pakistan's Water and Power Development Authority sought expressions of interest for the *Tarbela-V Extension Project* in August 2014, and was given final consent for construction in September 2015.

Construction is yet to begin, but will require an estimated 3.5 years for completion. The project will require installation of three turbines with a capacity of 470 MW each in Tarbela's fifth tunnel which was previously dedicated to agricultural use. Upon completion, the total power generating capacity of Tarbela Dam will be increased to 6,298 MW.

The hydro power project of tunnel no. 5 has two major components: power generation facilities and power evacuation facilities. The major works included under the project are: modifications to tunnel no. 5 and building a new power house and its ancillaries to generate about 1,800GWh of power annually, a new 50 km of 500kV double-circuit transmission line from Tarbela to the Islamabad West Grid Station for power evacuation, and a new 500kV Islamabad West Grid Station.

Financing

In November 2015, the World Bank affirmed that it would finance at least $326 million of the project's estimated $796 million cost which includes $126 million of funding that was diverted from the $840 million fourth phase extension project after costs for that project were revised downwards. On September 2016, the World Bank approved an additional financing of $390 million for the fifth extension hydropower project of Tarbela dam that will support the scaling up of the power generation capacity by adding 1,410 megawatts to the existing tunnel no. 5 of the dam. The project will be financed by the International Bank for Reconstruction and Development (IBRD), with a variable spread and 20-year maturity, including a six-year grace period. This will be the first World Bank-supported project in South Asia to be jointly financed with the Asian Infrastructure Investment Bank (AIIB) which will be providing $300m and the Government of Pakistan $133.5m. The total cost of the project is $823.5m.

Boguchany Dam

The Boguchany Dam is a large hydroelectric dam on the Angara River in Kodinsk, Krasnoyarsk Krai, Russia. It has an installed capacity of 2,997 MW. Construction of the power plant was completed when a ninth and final generator was brought online in January 2015.

History

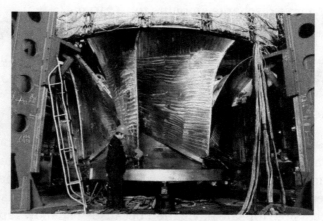

One of the nine turbines of the dam.

Preparatory works for construction started in 1974. The design was performed by Hydroproject in 1976. Construction of the power station started in 1980 but was suspended in 1994 due to the lack of financing. Work on the project resumed in 2005 when RAO UES (then owner of RusHydro) and Rusal agreed to develop the project jointly. Construction restarted in 2007. The first turbine was dispatched in 2008. The dam began to fill its reservoir in May 2012 and by August of that year, the first two turbines were installed. They later became operational on 15 October 2012. The third generator began operations later the same month and the fourth was commissioned on 13 January 2013.

Description

The Boguchany Dam is a 2,587 m (8,488 ft) long dam of combined type, which includes reinforced concrete 774 m (2,539 ft) long gravity segment for power station building and 1,813 m (5,948 ft) long rock-fill segment with asphaltene-concrete diaphragm throughout all its length. The dam was equipped with ship lock until 2010, later it was closed and its former location was included in concrete gravity section.

The power station consists of nine Francis turbines with a capacity of 333 MW each. It generates 17.6 TWh of electricity per year. Turbines are manufactured by Power Machines. The power station is owned and operated by JSC Boguchanskaya GES, a joint venture of RusHydro and Rusal, who also built it. Energy from the plant is partly used by the Boguchany Aluminium Smelter.

Locale

Main facilities of the dam are located on the Angara river, in Kodinsk gorge, 444 km upstream from the river mouth, and are surrounded by taiga. The Angara flows in latitudinal direction here, cutting through Cambrian and Ordovician sedimentary massif interspersed with diabase intrusions. The valley's width barely exceeds 1 mile here, with rocky asymmetric bluffs overhanging the riverstream. The region is potentially seismic, with up to 7 degrees MSK-64 earthquake chance (1 in 5000 years possible recurrence).

As it reaches the dam, The Angara's river basin area accounts for 831,000 km3. The river's supply is greatly dependent on the lake Baikal and superincumbent reservoirs of Irkutsk, Bratsk and Ust-Ilimsk dams. The Angara's discharge amounts to 16,210 m³/s.

The climatic conditions in the vicinity may be described as continental. Summers are short and warm, whilst winters are prolonged and severe. Annual average temperature ranges between −2,6 to −4,3 °C, with the average July high of +18,5 °C and the average January low of −27,4 °C.

Navigation Conditions Improvement

In 2012, the present design of all dams on the Angara river doesn't allow passage of any ships. Additionally the river was not navigable within the section between Boguchany Dam and Ust-Ilimsk Dam. After its completion in 2012-2013 the reservoir of the Boguchany Dam will be 375 km (233 mi) long and will reach Ust-Ilimsk Dam upstream. It will flood the last section of the Angara channel, which was inaccessible for safe navigation because of shallow rapids in the river.

The downstream part of the river has a shallow channel too and stays unavailable for large seaworthy ships. According to various sources related to this subject, there are plans to build additional dams in the lower part of Angara river or, according to original soviet plans, to build one large hydro power plant in Yenisei River below its join with Angara. The latter variant presumes construction of the largest hydo power plant in Russia with an annual production of over 40 TWh.

Atatürk Dam

The Atatürk Dam (Turkish: *Atatürk Barajı*), originally the Karababa Dam, is a zoned rock-fill dam with a central core on the Euphrates River on the border of Adıyaman Province and Şanlıurfa Province in the Southeastern Anatolia Region of Turkey. Built both to generate electricity and to irrigate the plains in the region, it was renamed in honour of Mustafa Kemal Atatürk (1881–1938), the founder of the Turkish Republic. The construction began in 1983 and was completed in 1990. The dam and the hydroelectric power plant, which went into service after the upfilling of the reservoir was completed in 1992, are operated by the State Hydraulic Works (DSİ). The reservoir created behind the dam, called Lake Atatürk Dam (Turkish: *Atatürk Baraj Gölü*), is the third largest in Turkey.

The dam is situated 23 km (14 mi) northwest of Bozova, Şanlıurfa Province, on state road D-875 from Bozova to Adıyaman. Centerpiece of the 22 dams on the Euphrates and the Tigris, which comprise the integrated, multi-sector, Southeastern Anatolia Project (Turkish: *Güney Doğu Anadolu Projesi*, known as GAP), it is one of the world's largest dams. The Atatürk Dam, one of the five operational dams on the Euphrates as of 2008, was preceded by Keban and Karakaya dams upstream and followed by Birecik and the Karkamış dams downstream. Two more dams on the river have been under construction.

The dam embankment is 169 m high (554 ft) and 1,820 m long (5,970 ft). The hydroelectric power plant (HEPP) has a total installed power capacity of 2,400 MW and generates 8,900 GW·h electricity annually. The total cost of the dam project was about US$1,250,000,000.

The dam was depicted on the reverse of the Turkish one-million-lira banknotes of 1995–2005 and of the 1 new lira banknote of 2005–2009.

Dam

The initial development project for the southeastern region of Turkey was presented in 1970. As the objectives for regional development have changed significantly and the ambitions have grown in the 1970s, the original plan underwent major modifications. The most important change in the project was abandoning the Middle Karababa Dam design, and adopting the design of the Atatürk Dam to increase the storage and power generation capacities of the dam.

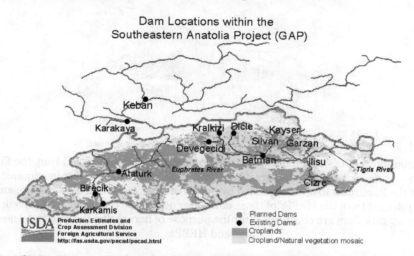

Dolsar Engineering and ATA Construction, two prominent Turkish companies, signed for the building of the dam. The construction of the cofferdam began in 1985 and was completed in 1987. The fill work for the main dam lasted from 1987 to 1990. The Atatürk Dam, listed in international construction publications as the world's largest construction site, was completed in a world record time of around 50 months.

The rock-fill dam undergoes deformations that are regularly and systematically monitored since 1990 with different types of sensors. It is estimated that the central portion of the dam crest has settled by around 7 m (23 ft) since the end of the construction. Settlement of the dam crest up to 4.3 m (14 ft) has been measured since the start of the detailed geodetic monitoring in 1992. The maximum horizontal (radial) deformation measured is about 2.9 m (9.5 ft).

The permeation grouting work was carried out by subcontractor Solétanche Bachy and the rehabilitation work for the post-tensioning of the dam crest with ground anchors by Vorspann System Losinger International (VSL).

Hydroelectric Power Plant

The HEPP of the Atatürk Dam is the biggest of a series of 19 power plants of the GAP project. It consists of eight Francis turbine and generator groups of 300 MW each, supplied by Sulzer Escher Wyss and ABB Asea Brown Boveri respectively. The up to 7,25 m dia steel pressure pipes (penstocks) with a total weight of 26.600 tons were supplied and installed by the German NOELL company (today DSD NOELL). The power plant's first two power units came on line in 1992 and it became fully operational in December 1993. The HEPP can generate 8,900 GWh of electricity annually. Its capacity makes up around one third of the total capacity of the GAP project.

During the periods of low demand for electricity, only one of the eight units of the HEPP is in operation while in times of high demand, all the eight units are in operation. Hence, depending upon the energy demand and the state of the interconnected system, the amount of water to be released from the HEPP might vary between 200 and 2,000 m³/s in one day.

Irrigation

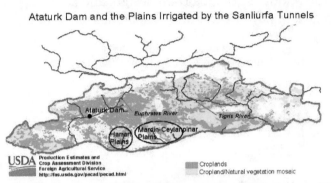

Ataturk Dam and the Plains Irrigated by the Sanliurfa Tunnels

Originating in the mountains of eastern Anatolia and flowing southwards to Syria and Iraq, the Euphrates and the Tigris are very irregular rivers, used to cause great problems each year with droughts in summer and flooding in winter. The water of the Euphrates River is regulated by means of large reservoirs of the Keban and Atatürk Dams. However, the waters released from the HEPPs of those dams also need to be regulated. The Birecik and the Karkamış Dams downstream the Atatürk Dam are constructed for the purpose of harnessing the waters released from large-scale dams and HEPPs.

Nearly 4,760 km² (1,840 sq mi) of arable land in the Şanlıurfa-Harran and Mardin-Ceylanpınar plains in upper Mesopotamia is being irrigated via gravity-flow with water diverted from the Atatürk Dam through the Şanlıurfa Tunnels system, which consists of two parallel tunnels, each 26.4 km (16.4 mi) long and 7.62 m (25.0 ft) in diameter. The flow rate of water through the tunnels is about 328 m³/s (11,600 cu ft/s), which makes one-third of the total flow of the Euphrates. The tunnels are the largest in the world, in terms of length and flow rate, built for irrigation purposes. The first tunnel was completed in 1995 and the other in 1996. The reservoir behind the dam will irrigate another 406,000 ha by pumping for a total of 882,000 ha.

The Atatürk Dam and the Şanlıurfa Tunnel system are two major components of the GAP project. Irrigation started in the Harran Plain in the spring of 1995. The impact of the irrigation on the economy of the region is significant. In ninety percent of the irrigated area, cotton is planted. Irrigation expansion within the Harran plains also increased Southeastern Anatolia's cotton production from 164,000 to 400,000 metric tons in 2001, or nearly sixty percent. With almost 50% share of the country's cotton production, the region developed to the leader in Turkey.

Reservoir Lake

Reverse of the 1 million Turkish lira banknote, depicting the Atatürk Dam (1995–2005).

The reservoir Lake Atatürk Dam, extending over an area of 817 km² (315 sq mi) with a water volume of 48.7 km³ (63,400 million cu yd), ranks third in size in Turkey after Lake Van and Lake Tuz. The reservoir water level touched 535 m (1,755 ft) amsl in 1994. Since then, it varies between 526 and 537 m amsl. The full reservoir level is 542 m (1,778 ft), and the minimum operation level is 526 m (1,726 ft) amsl.

Some 10 towns and 156 villages of three provinces are located around the Lake Atatürk Dam. The lake provides a fisheries and recreation site. For transportation purposes, several ferries have been operated in the reservoir. The reservoir lake is called "sea" by local people.

Fishery

Atatürk Dam Lake is an abundant source of food for local people and also provides opportunities for recreational fishing. In 1992, around 200,000 young fish (fingerlings), propagated in DSI's Atatürk Fish Hatchery, were introduced into the reservoir. Since then, the figure of fingerlings stocked into the lake reached around 33 million.

Commercially fishing in the reservoir developed to a catch of around annually 1,000 tons of some fish species with a market value of US$1.26 million. 8 of the 12 fish species being caught are economically valuable. In addition, the lake has a potential for cage culture of 7,000 tons/year worth of US$14 million.

With the aim of utilizing the fishing potential and creating jobs for the lakeside populations, the reservoir is zoned to 21 fishing sectors, each one having a water products cooperative. Considering all aspects of fishery activities, the reservoir contributes in total US$15 million to GNP and generates employment for 1,600 people.

Recreation and Sports

In order to open the region to tourism, to introduce modern sports to the local people and to integrate the social and economic progress taking place in the region with sports as a drive, a water sports festival was established in 1994, which takes place each year in September. The young people in the region developed an interest in water sports and started to take part in international contests in the branches of sailing, rowing-canoeing, swimming and diving on the Lake Atatürk Dam.

Furthermore, International Atatürk Dam Sailing Competition takes place every year in October on the lake.

Resettlement and Salvaging Cultural Heritage

With the forming of the reservoir lake, more than a hundred hamlets and villages were inundated and about 55,000 people were forced to relocate, many of them resettling in nearby communities. According to other sources, the construction of the dam resulted in involuntary resettlement of between 45,000 and 53,500 people.

In 1989, the old town of Samosata (Samsat), capital of the ancient Commagene kingdom located in Adıyaman Province was flooded behind the Atatürk Dam. A new town with the same name Samsat was founded for the around 2,000 people dislocated.

The birthplace of the Ancient Greek poet Lucian was lost when the dam was created.

Since the entire GAP area was home to early civilizations of Hittites, Assyrians, Medes, Persians, Greeks, Romans, Arabs and Turks, and therefore rich in terms of historical remains, cultural heritage of the region was a concern. The subject of salvaging cultural heritages gained importance, particularly after inundation of Samsat.

The early Neolithic settlement of Nevalı Çori, site of some of the world's most ancient known temples and monumental sculpture, was discovered during rescue excavations before the dam was completed. Nevalı Çori was inundated by Atatürk Dam's reservoir.

Political Controversy

About 90% of Euphrates' total annual flow originates in Turkey, while the remaining part is added in Syria, but nothing is contributed further downstream in Iraq. In general, the stream varies greatly in its flow from season to season and year to year. As an example, the annual flow at the border with Syria ranged from 15.3 km³ (3.7 cu mi) in 1961 to 42.7 km³ (10.2 cu mi) in 1963.

One of the most important legal texts on the waters of the Euphrates-Tigris river system is the protocol annexed to the 1946 Treaty of Friendship and Good Neighborly Relations between Iraq and Turkey. The protocol provided the control and management of the Euphrates and the Tigris depending to a large extent on the regulations of flow in Turkish source areas. Turkey agreed to begin monitoring the two border-crossing rivers and to share related data with Iraq. In 1980, Turkey and Iraq further specified the nature of the earlier protocol by forming a joint committee on technical issues, which Syria joined later in 1982 as well. Turkey unilaterally guaranteed to allow 15.75 km³/year (500 m³/s) of water across the border to Syria without any formal agreement on the sharing of the Euphrates water.

Mid-January 1990, when the first phase of the dam was completed, Turkey held back the flow of the Euphrates entirely for a month to begin filling up the reservoir. Turkey had notified Syria and Iraq by November 1989 of her decision to fill the reservoir over a period of one month explaining the technical reasons and providing a detailed program for making up for the losses. The downstream neighbors protested vehemently. At this point, the Atatürk Dam has cut the flow from the Euphrates by about a third.

Syria and Iraq claim to be suffering severe water shortages due to the GAP development. Both countries allege that Turkey is intentionally withholding supplies from its downstream neighbors, turning water into a weapon. Turkey denies these claims, and insists it has always supplied its southern neighbors with the promised minimum of 500 m³/s (18,000 cu ft/s). It argues that Iraq and Syria in fact benefit from the regulated water by the dams as they protect all three riparian countries from seasonal droughts and floods.

Bakun Dam

The Bakun Dam is an embankment dam located in Sarawak, Malaysia, on the Balui River, a tributary or source of the Rajang River and some sixty kilometres west of Belaga. As part of the project, the second tallest concrete-faced rockfill dam in the world would be built. It is planned to generate 2,400 megawatts (MW) of electricity once completed.

The purpose for the dam was to meet growing demand for electricity. However, most of this demand is said to lie in Peninsular Malaysia and not East Malaysia, where the dam is located. Even in Peninsular Malaysia, however, there is an oversupply of electricity, with Tenaga Nasional Berhad being locked into unfavourable purchasing agreements with Independent Power Producers. The original idea was to have 30% of the generated capacity consumed in East Malaysia and the rest sent to Peninsular Malaysia. This plan envisioned 730 km of overhead HVDC transmission lines in East Malaysia, 670 km of undersea HVDC cable and 300 km of HVDC transmission line in Peninsular Malaysia.

Future plans for the dam include connecting it to an envisioned Trans-Borneo Power Grid Interconnection, which would be a grid to supply power to Sarawak, Sabah, Brunei, and Kalimantan (Indonesia). There have been mentions of this grid made within ASEAN meetings but no actions have been taken by any party. Bakun Dam came online on 6 August 2011.

As of 2015, Bakun Dam is the biggest dam in Southeast Asia.

Project History

First Attempt

Initial survey was conducted in the early 1960s and more studies were conducted in the early 1980s. The studies cover the masterplan and feasibility report, rock and soil studies, hydro potential, detailed design and costing, environmental and socio-economic studies and HVDC transmission studies. Notable consultants involved were SAMA Consortium German Agency for Technical Cooperation, Snowy Mountains Engineering Corporation and Maeda-Okumura Joint Venture, Fichtner and Swedpower Swedish Agency for Technical Cooperation.

Although the project was first approved by government in 1986, it was shelved in 1990 owing to decreased projection of electricity demand due to the recession of 1985 and the decision to use the then low-cost alternative of natural gas as fuel for developing the petrochemical industry.

Second Attempt

It was revived in September 1993 by the Malaysian Federal Government led by then-Prime Minister Mahathir Mohammad. In January 1994, a privatised contract was awarded to Ekran Berhad. In April 1995, Ekran completed the EIA of the project. The project was to cost US$2.4 billion and was originally scheduled for completion in 2003.

The dam was to be built beginning in 1994 by a privatised joint-venture consortium called Bakun Hydroelectric Corporation, comprising Ekran Berhad, Tenaga Nasional Berhad (TNB), the government of Sarawak, Sarawak Electricity Supply Corporation (Sesco), and Malaysia Mining Corporation Bhd (MMC).

Ekran awarded the electromechanical works and the transmission portion to ABB. ABB's consortium partner for the civil works will be Companhia Brasileira de Projetos e Obras (CBPO) of Brazil, a large civil engineering company belonging to the Odebrecht Group, responsible for the construction of the dam and power house. Engineering consulting firms involved in the project then were TNB Hydro, a subsidiary of Tenaga Nasional Berhad and KLIA Consult.

Ekran launched a rights issue to finance the building of the dam, but it was undersubscribed and Ting Pek Khing (Ekran's chairman) had to put up $500 million to take up the unsubscribed portion as part of his agreement with the underwriters. Ekran was a company of Ting Pek Khing, himself a timber businessman. Neither he nor his company had built a dam before. The entire project was not tendered publicly, and instead was awarded by government contract.

The project was halted in 1997 in the face of the Asian financial crisis. When the project was shelved, the Malaysian government took back the project from this consortium. By this time, RM1.6 billion had already been paid out by the government. RM700 million to RM1.1 billion was paid as 'compensation' to Ekran, according to figures disclosed in Parliament. The completed works were the river diversion tunnels by Dong Ah of Korea for RM400 million, and to Global Upline for work completed on the auxiliary cofferdams for RM60 million. Other works are for selective clearing of biomass, and relocation of the affected native residents. The government had also turned over RM1 billion for the purchase of eight turbines.

Third Attempt

In May 2000 it was revived through a 100% government-owned company, Sarawak Hidro, but the transmission of power to Peninsular Malaysia was not part of the revived project. The construction work was tendered out as a turnkey contract. The completion date has been revised to February 2008.

The new civil builder is the Malaysia–China Hydro JV consortium, led by Sime Engineering Berhad of Malaysia (a subsidiary of Sime Darby and Sinohydro Corporation of China. Other members of the consortium are WCT Berhad, MTD Capital, Ahmad Zaki Resources, Syarikat Ismail and Edward & Sons. It targeted a completion date of September 2007. The total sum to be paid to this consortium was budgeted at RM1.8 billion. The electromechanical works for the turbines were awarded in two contracts to IMPSA of Argentina and Alstom of France.

In 2004, engineering consulting firm JR Knowles, was hired to study the delays in construction. Other engineering consulting firms involved in the project were Snowy Mountain Engineering Corporation, of Australia and Opus International Malaysia.

Items of Temporary Interest During Third Attempt

In May 2004 Ting Pek Khing's name again was raised in connection with the project. A Ting-owned company, Global Upline, was rumoured to have been awarded a contract to undertake "biomass removal" in the flood basin. This would allow him to harvest timber in the area without a separate permit. Issuance of timber permits has come under increased scrutiny due to political conditions and environmental concerns. However, as of December 2006 it has not been awarded.

Usage of the generated capacity was to have been by a proposed aluminium smelting plant in Similajau, near Bintulu, approximately 180 km inland from the dam. The project is a joint venture between Dubai Aluminum Co, Ltd (Dubal) and Gulf International Investment Group (GIIG), an investment fund jointly set up by Malaysian tycoon Syed Mokhtar, and Dubai-based international financier Mohamed Ali Alabbar. This plant was expected to consume 50% of the power generated. The government has agreed in principle that 60% of Sarawak Hidro, the entity that owns the dam, will be sold to GIIG. Owing to delays in dam construction, the plans for the smelter have since been shelved. The agreement for this smelter was originally signed in 2003 and some conditions have lapsed owing to delays in construction. Rio Tinto announced in August 2007 that they had signed a deal with Malaysian conglomerate Cahya Mata Sarawak Berhad (CMSB) to build an aluminium smelter. The production capacity would be 550,000 tonnes initially with expansion to 1.5 million tonnes possible. Production of aluminium would start at the end of 2010.

At the end of 2004, the minor partners in the Malaysia–China Hydro JV consortium (Ahmad Zaki Resources Bhd, WCT Engineering Bhd and MTD Capital) will report quarterly losses due to the Bakun project. Discounting this project, they would all be operating profitably for the quarter.

- For Ahmad Zaki Resources Bhd, this is the first time it will report losses since 1993. Ahmad Zaki estimates net losses suffered for nine months ending 30 September 2004 at RM4.55 million.

- WCT has reported a net loss of RM13.08 million for the quarter ending 30 September 2004 due to the Bakun project.

- MTD Capital reported a RM4.04 million loss in the quarter ending 30 September 2004 and attributed it to "a major project undertaken by the company in joint venture with both local and foreign partners" without specifically naming Bakun.

- The lead partner in the project, Sime Engineering reported a profit of RM521,000 for the same quarter versus a profit of RM4.18 million for the same quarter in 2003. Declines in profit were attributed to cost overruns and project delays.

As of February 2007, there are three developments affecting the Bakun project. The first is the merger of the Sime Darby, Guthrie and Golden Hope into a new entity named Synergy Drive. The second is the proposed takeover of the Bakun project by the contractor, Sime Engineering. The third is the revival of the submarine HVDC cable under the South China Sea to transmit electricity from Borneo to Peninsular Malaysia.

In November 2007, Sime Darby, the parent company of the contractor Sime Engineering will be merged with Golden Hope and Guthrie into a new company with a market capitalisation of RM31 billion (US$8 billion). Concurrently with the merger, the contractor Sime Engineering will takeover the ownership of the Bakun Dam project.

Sime Engineering Sdn Bhd has filed a suit against AZRB over alleged breaches in the Malaysia–China Hydro joint venture agreement dated 12 June 2002 relating to the Bakun dam. AZRB was served with a writ summons and statement of claim dated 12 Oct by Sime Engineering claiming "RM15.24 million for alleged breaches by AZRB of the Malaysia–China Hydro JVA" relating to Bakun hydroelectric project package CW2 – main civil works.

Sinohydro Flawed Construction Procedures

Chinese constructor Sinohydro has acknowledged that its construction procedures used for Bakun were flawed. The admission came after *Sarawak Report* reported that Sinohydro had widely used a technique involving adding excessive water to cement, with potentially dangerous consequences. The *Sarawak Report* said its website was attacked after it published the report.

Environmental and Social Damage

Uma Daro longhouse is one of the longhouses built in Sungai Asap for the relocation of the natives displaced by the Bakun Dam.

The Bakun dam flooding commenced on 13 October 2010 with a faulty start and will put 700 km² of land underwater – equivalent to the size of Singapore. The rainforest of this part of Southeast Asia has some of the highest rates of plant and animal endemism, species found there and nowhere else on Earth, and this dam has done irreparable ecological damage to that region.

Construction of the dam required the relocation of more than 9,000 native residents (mainly Kayan/Kenyah) of the indigenous peoples who lived in the area to be flooded. Many Sarawak natives have been relocated to a longhouse settlement named Sungai Asap in Bakun. Most of them were subsistence farmers. Each family was promised 3 acres (1.21 ha) of land but many families still have not been compensated.

Concerns were raised also about such things as the relocation of people; the amount of virgin tropical rainforest that had to be cut down (230 km²); possible dam collapse issues; increase in diseases with waterborne vectors such as schistosomiasis, opisthorchiasis, malaria, and filariasis; and sediment accumulation shortening the useful lifespan of the dam. A 5-part series of Bakun dam documentaries was filmed by Chou Z Lam. The series highlighted the basic community problems faced by displaced indigenous people such as the lack of land areas for farming and hunting, lack of educational, medical, and transport facilities and also the promises not being kept by the government. This documentary series was later banned from Radio Television Malaysia (RTM) on May 2010, forcing the remaining series to YouTube.

Transparency International includes Bakun Dam in its 'Monuments of corruption' Global Corruption Report 2005. The mandate to develop the project went to a timber contractor and friend of Sarawak's governor. The provincial government of Sarawak is still looking for customers to consume the power to be generated by the project.

Launched in February 2012, an international NGO coalition that includes organisations from the US, Norway and Switzerland is showing its solidarity with Malaysian groups who are protesting against the construction of twelve hydroelectric dams in the Malaysian state of Sarawak on Borneo. The NGO coalition supports the Malaysian groups' demand for an immediate halt to the realisation of these dams, which threaten to displace tens of thousands of Sarawak natives and flood hundreds of square miles of Sarawak's precious tropical rainforests.

Technical Specifications

View of turbines inside the powerhouse in 2009

The permanent dam components are as follows:

- Main dam

  - Maximum height above foundation of 205 m, and crest length of 750 m, volume of fill is 16,710,000 m³.

  - Crest elevation is 235 m above sea level (ASL), maximum flood level is 232 m, operating levels maximum 228 m and minimum 195 m.

  - Reservoir area at 228 m ASL is 695 km², and with a catchment area of 14,750 km². Gross storage volume is 43,800 million cubic meters.

- Power intake structure – 8 bays with 16 roller gates.

- Gated spillway – gated concrete weir with chute and flip bucket, with capacity of 15,000 cubic meters per second.

- Power tunnels – 8 tunnels of 8.5 m diameter each with lengths of 760 m each.

- Powerhouse

  - Surface powerhouse, with 4-level measuring 250 m length × 48 m width × 48 m height

  - with 8 penstocks to powertrains comprising 8 vertical-shaft Francis turbines of 300 MW each, 8 air-cooled generators of 360 MVA each and 8 oil-immersed transformers of 360 MVA each.

Transmission Lines

There are four major transmission line sections:

The first consists of an HVAC double circuit overhead line running over a distance of 160 km from Bakun Dam to Similajau Static Inverter Plant, situated east of Bintulu.

The three further sections consist of a bipolar HVDC 500 kV line. The first section of this line running from Similajau Static Inverter Plant to Kampung Pueh on Borneo will be implemented as overhead lines with a length of 670 km.

The next section is the submarine cable between Kampung Pueh to Tanjung Leman, Johor. It will have a length of 670 km. It is planned to be implemented by 3 or 4 parallel cables each with a transmission capacity of 700 MW.

The last section on the Malay Peninsula will consist of an overhead DC powerline running from Tanjung Leman to the static inverter plant at Bentong.

As part of the transmission works two converter stations will be built at Bakun and Tanjung Tenggara. The HVDC lines will connect to the National Grid, Malaysia operated by Tenaga Nasional Berhad.

Revival of Submarine Cable Component

The revived submarine cable portion is to transmit the electricity generated at Bakun Dam in Borneo to Peninsular Malaysia, possibly by 2012. The consortium partners' equity possibly will be Sime Darby (60%), Tenaga Nasional (20%) and the Malaysian Ministry of Finance (20%). The consortium is exploring financing facilities of up to 80% of the planned investment.

The cable is planned to transmit 1,600 MW of power from the Bakun Dam to Yong Peng, Johor by undersea HVDC power cables and then by land line onto the Malaysian National Grid. The use of HVDC cables would ensure that the energy loss is minimal, at about 5% to 6% only. The cost of the undersea cable is estimated at RM9 billion. The proposed concept is for two 800 MW cables being laid about 660 km under the South China Sea from the Sarawak shore to Yong Peng on Peninsular Malaysia.

Sime Darby would take ownership of the submarine cable project but not undertake its construction. The contractor is rumoured to be Malaysian Resources Corporation Berhad (MRCB), a public listed company on the KLSE.

The buyer of the electricity is Tenaga Nasional. The rate proposed is RM0.17 per kilowatt hour at the intake onto the National Grid. Analysts estimated that generation cost using world-market-rate natural gas would cost RM0.22 per kilowatt hour. A 4% increase every 4 years is envisaged over the 35-year concession period.

On 7 January 2008, Sime Darby announced that they had appointed a financial adviser for the undersea power transmission project. However, the company did not name the financial adviser.

After many delays, Sarawak Energy Berhad announced that the contract to build the submarine cable would be awarded in mid-2010 with international tenders to be called in early 2010. It was expected that the construction would be completed by 2015 at an estimated cost of MYR8 billion to MYR10 billion. However, the project has been shelved.

Records

Once completed:

- Bakun Dam will be the tallest concrete-faced rockfill dam (CFRD) in the world.

- Bakun Lake will be the biggest lake in Malaysia by storage volume.

- Bakun Lake will be the largest lake in Malaysia by surface area, even though it is not apparent on the map, owing to the narrow shapes of the various lake arms, as a result of its location in the highland valleys.

- Bakun Power Station will be the largest hydroelectric dam in Malaysia, surpassing the currently largest Pergau Dam.

- Bakun submarine power cable will be the longest in the world, surpassing the current Norway-to-Netherlands submarine cable.

Hoover Dam

Hoover Dam, once known as Boulder Dam, is a concrete arch-gravity dam in the Black Canyon of the Colorado River, on the border between the U.S. states of Nevada and Arizona. It was constructed between 1931 and 1936 during the Great Depression and was dedicated on September 30, 1935, by President Franklin D. Roosevelt. Its construction was the result of a massive effort involving thousands of workers, and cost over one hundred lives. The dam was controversially named after President Herbert Hoover.

Since about 1900, the Black Canyon and nearby Boulder Canyon had been investigated for their potential to support a dam that would control floods, provide irrigation water and produce hydroelectric power. In 1928, Congress authorized the project. The winning bid to build the dam was submitted by a consortium called Six Companies, Inc., which began construction on the dam in early 1931. Such a large concrete structure had never been built before, and some of the techniques were unproven. The torrid summer weather and lack of facilities near the site also presented difficulties. Nevertheless, Six Companies turned over the dam to the federal government on March 1, 1936, more than two years ahead of schedule.

Hoover Dam impounds Lake Mead, the largest reservoir in the United States by volume (when it is full). The dam is located near Boulder City, Nevada, a municipality originally constructed for workers on the construction project, about 30 mi (48 km) southeast of Las Vegas, Nevada. The dam's generators provide power for public and private utilities in Nevada, Arizona, and California. Hoover Dam is a major tourist attraction; nearly a million people tour the dam each year. The heavily travelled U.S. 93 ran along the dam's crest until October 2010, when the Hoover Dam Bypass opened.

Background

Search for Resources

As the United States developed the Southwest, the Colorado River was seen as a potential source of irrigation water. An initial attempt at diverting the river for irrigation purposes occurred in the late 1890s, when land speculator William Beatty built the Alamo Canal just north of the Mexican border; the canal dipped into Mexico before running to a desolate area Beatty named the Imperial

Valley. Though water from the Imperial Canal allowed for the widespread settlement of the valley, the canal proved expensive to maintain. After a catastrophic breach that caused the Colorado River to fill the Salton Sea, the Southern Pacific Railroad spent $3 million in 1906–07 to stabilize the waterway, an amount it hoped vainly would be reimbursed by the Federal Government. Even after the waterway was stabilized, it proved unsatisfactory because of constant disputes with landowners on the Mexican side of the border.

River view of the future site of Hoover Dam, circa 1904

As the technology of electric power transmission improved, the Lower Colorado was considered for its hydroelectric-power potential. In 1902, the Edison Electric Company of Los Angeles surveyed the river in the hope of building a 40-foot (12 m) rock dam which could generate 10,000 horsepower (7,500 kW). However, at the time, the limit of transmission of electric power was 80 miles (130 km), and there were few customers (mostly mines) within that limit. Edison allowed land options it held on the river to lapse—including an option for what became the site of Hoover Dam.

In the following years, the Bureau of Reclamation (BOR), known as the Reclamation Service at the time, also considered the Lower Colorado as the site for a dam. Service chief Arthur Powell Davis proposed using dynamite to collapse the walls of Boulder Canyon, 20 miles (32 km) north of the eventual dam site, into the river. The river would carry off the smaller pieces of debris, and a dam would be built incorporating the remaining rubble. In 1922, after considering it for several years, the Reclamation Service finally rejected the proposal, citing doubts about the unproven technique and questions as to whether it would in fact save money.

Planning and Agreements

Sketch of proposed Boulder Canyon dam site and reservoir, circa 1921

In 1922, the Reclamation Service presented a report calling for the development of a dam on the

Colorado River for flood control and electric power generation. The report was principally authored by Davis, and was called the Fall-Davis report after Interior Secretary Albert Fall. The Fall-Davis report cited use of the Colorado River as a federal concern, because the river's basin covered several states, and the river eventually entered Mexico. Though the Fall-Davis report called for a dam "at or near Boulder Canyon", the Reclamation Service (which was renamed the Bureau of Reclamation the following year) found that canyon unsuitable. One potential site at Boulder Canyon was bisected by a geologic fault; two others were so narrow there was no space for a construction camp at the bottom of the canyon or for a spillway. The Service investigated Black Canyon and found it ideal; a railway could be laid from the railhead in Las Vegas to the top of the dam site. Despite the site change, the dam project was referred to as the "Boulder Canyon Project".

With little guidance on water allocation from the Supreme Court, proponents of the dam feared endless litigation. A Colorado attorney proposed that the seven states which fell within the river's basin (California, Nevada, Arizona, Utah, New Mexico, Colorado and Wyoming) form an interstate compact, with the approval of Congress. Such compacts were authorized by Article I of the United States Constitution but had never been concluded among more than two states. In 1922, representatives of seven states met with then-Secretary of Commerce Herbert Hoover. Initial talks produced no result, but when the Supreme Court handed down the *Wyoming v. Colorado* decision undermining the claims of the upstream states, they became anxious to reach an agreement. The resulting Colorado River Compact was signed on November 24, 1922.

Legislation to authorize the dam was introduced repeatedly by Representative Phil Swing (R-Calif.) and Senator Hiram Johnson (R-Calif.), but representatives from other parts of the country considered the project as hugely expensive and one that would mostly benefit California. The 1927 Mississippi flood made Midwestern and Southern congressmen and senators more sympathetic toward the dam project. On March 12, 1928, the failure of the St. Francis Dam, constructed by the city of Los Angeles, caused a disastrous flood that killed up to 600 people. As that dam was a curved-gravity type, similar in design to the arch-gravity as was proposed for the Black Canyon dam, opponents claimed that the Black Canyon dam's safety could not be guaranteed. Congress authorized a board of engineers to review plans for the proposed dam. The Colorado River Board found the project feasible, but warned that should the dam fail, every downstream Colorado River community would be destroyed, and that the river might change course and empty into the Salton Sea. The Board cautioned: "To avoid such possibilities, the proposed dam should be constructed on conservative if not ultra-conservative lines."

On December 21, 1928 President Coolidge signed the bill authorizing the dam. The Boulder Canyon Project Act appropriated $165 million for the Hoover Dam along with the downstream Imperial Dam and All-American Canal, a replacement for Beatty's canal entirely on the U.S. side of the border. It also permitted the compact to go into effect when at least six of the seven states approved it. This occurred on March 6, 1929 with Utah's ratification; Arizona did not approve it until 1944.

Design, Preparation and Contracting

Even before Congress approved the Boulder Canyon Project, the Bureau of Reclamation was considering what kind of dam should be used. Officials eventually decided on a massive concrete arch-gravity dam, the design of which was overseen by the Bureau's chief design engineer John L. Savage. The monolithic dam would be thick at the bottom and thin near the top, and would present a convex face towards the water above the dam. The curving arch of the dam would transmit the water's force into the abutments, in this case the rock walls of the canyon. The wedge-shaped dam would be 660 ft (200 m) thick at the bottom, narrowing to 45 ft (14 m) at the top, leaving room for a highway connecting Nevada and Arizona.

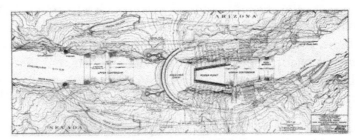
Hoover Dam architectural plans

On January 10, 1931, the Bureau made the bid documents available to interested parties, at five dollars a copy. The government was to provide the materials; but the contractor was to prepare the site and build the dam. The dam was described in minute detail, covering 100 pages of text and 76 drawings. A $2 million bid bond was to accompany each bid; the winner would have to post a $5 million performance bond. The contractor had seven years to build the dam, or penalties would ensue.

The Wattis Brothers, heads of the Utah Construction Company, were interested in bidding on the project, but lacked the money for the performance bond. They lacked sufficient resources even in combination with their longtime partners, Morrison-Knudsen, which employed the nation's leading dam builder, Frank Crowe. They formed a joint venture to bid for the project with Pacific Bridge Company of Portland, Oregon; Henry J. Kaiser & W. A. Bechtel Company of San Francisco; MacDonald & Kahn Ltd. of Los Angeles; and the J.F. Shea Company of Portland, Oregon. The joint venture was called Six Companies, Inc. as Bechtel and Kaiser were considered one company for purposes of 6 in the name. The name was descriptive and was an inside joke among the San Franciscans in the bid, where "Six Companies" was also a Chinese benevolent association in the city. There were three valid bids, and Six Companies' bid of $48,890,955 was the lowest, within $24,000 of the confidential government estimate of what the dam would cost to build, and five million dollars less than the next-lowest bid.

The city of Las Vegas had lobbied hard to be the headquarters for the dam construction, closing its many speakeasies when the decision maker, Secretary of the Interior Ray Wilbur came to town. Instead, Wilbur announced in early 1930 that a model city was to be built in the desert near the dam site. This town became known as Boulder City, Nevada. Construction of a rail line joining Las Vegas and the dam site began in September 1930.

Construction

Labor Force

"Apache Indians employed as high-scalers on the construction of Hoover Dam." - NARA

Soon after the dam was authorized, increasing numbers of unemployed people converged on southern Nevada. Las Vegas, then a small city of some 5,000, saw between 10,000 and 20,000 unemployed descend on it. A government camp was established for surveyors and other personnel near the dam site; this soon became surrounded by a squatters' camp. Known as McKeeversville, the camp was home to men hoping for work on the project, together with their families. Another camp, on the flats along the Colorado River, was officially called Williamsville, but was known to its inhabitants as "Ragtown". When construction began, Six Companies hired large numbers of workers, with more than 3,000 on the payroll by 1932 and with employment peaking at 5,251 in July 1934. "Mongolian" (Chinese) labor was prevented by the construction contract, while the number of blacks employed by Six Companies never exceeded thirty, mostly lowest-pay-scale laborers in a segregated crew, who were issued separate water buckets.

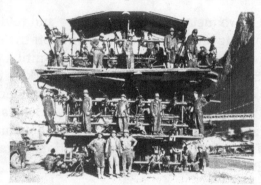

Workers on a "Jumbo Rig"; used for drilling Hoover Dam's tunnels

As part of the contract, Six Companies, Inc. was to build Boulder City to house the workers. The original timetable called for Boulder City to be built before the dam project began, but President Hoover ordered work on the dam to begin in March 1931 rather than in October. The company built bunkhouses, attached to the canyon wall, to house 480 single men at what became known as River Camp. Workers with families were left to provide their own accommodations until Boulder City could be completed, and many lived in Ragtown. The site of Hoover Dam endures extremely hot weather, and the summer of 1931 was especially torrid, with the daytime high averaging 119.9 °F (48.8 °C). Sixteen workers and other riverbank residents died of heat prostration between June 25 and July 26, 1931.

General Superintendent Frank Crowe (right) with Bureau of Reclamation engineer Walker Young in 1935

The Industrial Workers of the World (IWW or "Wobblies"), though much-reduced from their heyday as militant labor organizers in the early years of the century, hoped to unionize the Six Companies workers by capitalizing on their discontent. They sent eleven organizers, several of whom were arrested by Las Vegas police. On August 7, 1931, the company cut wages for all tunnel workers. Although the workers sent away the organizers, not wanting to be associated with the "Wobblies", they formed a committee to represent them with the company. The committee drew up a list of demands that evening and presented them to Crowe the following morning. He was noncommittal. The workers hoped that Crowe, the general superintendent of the job, would be sympathetic; instead he gave a scathing interview to a newspaper, describing the workers as "malcontents".

On the morning of the 9th, Crowe met with the committee and told them that management refused their demands, was stopping all work, and was laying off the entire work force, except for a few office workers and carpenters. The workers were given until 5 p.m. to vacate the premises. Concerned that a violent confrontation was imminent, most workers took their paychecks and left for Las Vegas to await developments. Two days later, the remainder were talked into leaving by law enforcement. On August 13, the company began hiring workers again, and two days later, the strike was called off. While the workers received none of their demands, the company guaranteed there would be no further reductions in wages. Living conditions began to improve as the first residents moved into Boulder City in late 1931.

A second labor action took place in July 1935, as construction on the dam wound down. When a Six Companies manager altered working times to force workers to take lunch on their own time, workers responded with a strike. Emboldened by Crowe's reversal of the lunch decree, workers raised their demands to include a $1-per-day raise. The company agreed to ask the Federal government to supplement the pay, but no money was forthcoming from Washington. The strike ended.

River Diversion

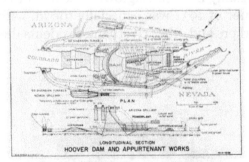

Overview of dam mechanisms; diversion tunnels shown

Before the dam could be built, the Colorado River needed to be diverted away from the construction site. To accomplish this, four diversion tunnels were driven through the canyon walls, two on the Nevada side and two on the Arizona side. These tunnels were 56 ft (17 m) in diameter. Their combined length was nearly 16,000 ft, or more than 3 miles (5 km). The contract required these tunnels to be completed by October 1, 1933, with a $3,000-per-day fine to be assessed for any delay. To meet the deadline, Six Companies had to complete work by early 1933, since only in late fall and winter was the water level in the river low enough to safely divert.

Tunneling began at the lower portals of the Nevada tunnels in May 1931. Shortly afterward, work began on two similar tunnels in the Arizona canyon wall. In March 1932, work began on lining the tunnels with concrete. First the base, or invert, was poured. Gantry cranes, running on rails through the entire length of each tunnel were used to place the concrete. The sidewalls were poured next. Movable sections of steel forms were used for the sidewalls. Finally, using pneumatic guns, the overheads were filled in. The concrete lining is 3 feet (1 m) thick, reducing the finished tunnel diameter to 50 ft (15 m). The river was diverted into the two Arizona tunnels on November 13, 1932; the Nevada tunnels were kept in reserve for high water. This was done by exploding a temporary cofferdam protecting the Arizona tunnels while at the same time dumping rubble into the river until its natural course was blocked.

Following the completion of the dam, the entrances to the two outer diversion tunnels were sealed at the opening and halfway through the tunnels with large concrete plugs. The downstream halves of the tunnels following the inner plugs are now the main bodies of the spillway tunnels. The inner diversion tunnels were plugged at approximately one-third of their length, beyond which they now carry steel pipes connecting the intake towers to the power plant and outlet works. The inner tunnels' outlets are equipped with gates that can be closed to drain the tunnels for maintenance.

Groundworks, Rock Clearance and Grout Curtain

To protect the construction site from the Colorado River and to facilitate the river's diversion, two cofferdams were constructed. Work on the upper cofferdam began in September 1932, even though the river had not yet been diverted. The cofferdams were designed to protect against the possibility of the river flooding a site at which two thousand men might be at work, and their specifications were covered in the bid documents in nearly as much detail as the dam itself. The upper cofferdam was 96 ft (29 m) high, and 750 feet (230 m) thick at its base, thicker than the dam itself. It contained 650,000 cubic yards (500,000 m³) of material.

Looking down at "high scalers" above the Colorado River

When the cofferdams were in place and the construction site was drained of water, excavation for the dam foundation began. For the dam to rest on solid rock, it was necessary to remove accumulated erosion soils and other loose materials in the riverbed until sound bedrock was reached. Work on the foundation excavations was completed in June 1933. During this excavation, approximately 1,500,000 cu yd (1,100,000 m³) of material was removed. Since the dam was an arch-gravity type, the side-walls of the canyon would bear the force of the impounded lake. Therefore, the side-walls were excavated too, to reach virgin rock as weathered rock might provide pathways for water seepage.

The men who removed this rock were called "high scalers". While suspended from the top of the canyon with ropes, the high-scalers climbed down the canyon walls and removed the loose rock with jackhammers and dynamite. Falling objects were the most common cause of death on the dam site; the high scalers' work thus helped ensure worker safety. One high scaler was able to save life in a more direct manner: when a government inspector lost his grip on a safety line and began tumbling down a slope towards almost certain death, a high scaler was able to intercept him and pull him into the air. The construction site had, even then, become a magnet for tourists; the high scalers were prime attractions and showed off for the watchers. The high scalers received considerable media attention, with one worker dubbed the "Human Pendulum" for swinging co-workers (and, at other times, cases of dynamite) across the canyon. To protect themselves against falling objects, some high scalers took cloth hats and dipped them in tar, allowing them to harden. When workers wearing such headgear were struck hard enough to inflict broken jaws, they sustained no skull damage, Six Companies ordered thousands of what initially were called "hard boiled hats" (later "hard hats") and strongly encouraged their use.

The cleared, underlying rock foundation of the dam site was reinforced with grout, called a grout curtain. Holes were driven into the walls and base of the canyon, as deep as 150 feet (46 m) into the rock, and any cavities encountered were to be filled with grout. This was done to stabilize the rock, to prevent water from seeping past the dam through the canyon rock, and to limit "uplift"—upward pressure from water seeping under the dam. The workers were under severe time constraints due to the beginning of the concrete pour, and when they encountered hot springs or cavities too large to readily fill, they moved on without resolving the problem. A total of 58 of the 393 holes were incompletely filled. After the dam was completed and the lake began to fill, large numbers of significant leaks into the dam caused the Bureau of Reclamation to look into the situation. It found that the work had been incompletely done, and was based on less than a full understanding of the canyon's geology. New holes were drilled from inspection galleries inside the dam into the surrounding bedrock. It took nine years (1938–47) under relative secrecy to complete the supplemental grout curtain.

Concrete

Columns of Hoover Dam being filled with concrete, February 1934 (looking upstream from the Nevada rim)

The first concrete was poured into the dam on June 6, 1933, 18 months ahead of schedule. Since concrete heats and contracts as it cures, the potential for uneven cooling and contraction of the concrete posed a serious problem. Bureau of Reclamation engineers calculated that if the dam was built in a single continuous pour, the concrete would take 125 years to cool, and the resulting stresses would cause the dam to crack and crumble. Instead, the ground where the dam would rise was marked with rectangles, and concrete blocks in columns were poured, some as large as 50 ft square (15 m) and 5 feet (1.5 m) high. Each five-foot form contained a series of 1-inch (25 mm) steel pipes; cool riverwater would be poured through the pipes, followed by ice-cold water from a refrigeration plant. When an individual block had cured and had stopped contracting, the pipes were filled with grout. Grout was also used to fill the hairline spaces between columns, which were grooved to increase the strength of the joins.

The concrete was delivered in huge steel buckets 7 feet high (2.1 m) and almost 7 feet in diameter; Crowe was awarded two patents for their design. These buckets, which weighed 20 short tons (18 t) when full, were filled at two massive concrete plants on the Nevada side, and were delivered to the site in special railcars. The buckets were then suspended from aerial cableways, which were used to deliver the bucket to a specific column. As the required grade of aggregate in the concrete differed depending on placement in the dam (from pea-sized gravel to 9-inch or 23 cm stones), it was vital that the bucket be maneuvered to the proper column. When the bottom of the bucket opened up, disgorging 8 cu yd (6.1 m³) of concrete, a team of men worked it throughout the form. Although there are myths that men were caught in the pour and are entombed in the dam to this day, each bucket only deepened the concrete in a form by an inch, and Six Companies engineers would not have permitted a flaw caused by the presence of a human body.

A total of 3,250,000 cubic yards (2,480,000 m³) of concrete was used in the dam before concrete pouring ceased on May 29, 1935. In addition, 1,110,000 cu yd (850,000 m³) were used in the power plant and other works. More than 582 miles (937 km) of cooling pipes were placed within the concrete. Overall, there is enough concrete in the dam to pave a two-lane highway from San Francisco to New York. Concrete cores were removed from the dam for testing in 1995; they showed that "Hoover Dam's concrete has continued to slowly gain strength" and the dam is composed of a "durable concrete having a compressive strength exceeding the range typically found in normal mass concrete". Hoover Dam concrete is not subject to alkali–silica reaction (ASR) as the Hoover Dam builders happened to use nonreactive aggregate, unlike that at downstream Parker Dam, where ASR has caused measurable deterioration.

Dedication and Completion

The upstream face of Hoover Dam slowly disappears as Lake Mead fills, May 1935 (looking downstream from the Arizona rim)

With most work finished on the dam itself (the powerhouse remained uncompleted), a formal dedication ceremony was arranged for September 30, 1935, to coincide with a western tour being made by President Franklin D. Roosevelt. The morning of the dedication, it was moved forward three hours from 2 p.m. Pacific time to 11 a.m.; this was done because Secretary of the Interior Harold L. Ickes had reserved a radio slot for the President for 2 p.m. but officials did not realize until the day of the ceremony that the slot was for 2 p.m. Eastern Time. Despite the change in the ceremony time, and temperatures of 102 °F (39 °C), 10,000 people were present for the President's speech, in which he avoided mentioning the name of former President Hoover, who was not invited to the ceremony. To mark the occasion, a three-cent stamp was issued by the United States Post Office Department—bearing the name "Boulder Dam", the official name of the dam between 1933 and 1947. After the ceremony, Roosevelt made the first visit by any American president to Las Vegas.

Most work had been completed by the dedication, and Six Companies negotiated with the government through late 1935, and early 1936, to settle all claims and arrange for the formal transfer of the dam to the Federal Government. The parties came to an agreement and on March 1, 1936, Secretary Ickes formally accepted the dam on behalf of the government. Six Companies was not required to complete work on one item, a concrete plug for one of the bypass tunnels, as the tunnel had to be used to take in irrigation water until the powerhouse went into operation.

Construction Deaths

Oskar J. W. Hansen's memorial at the dam which reads in part "They died to make the desert bloom."

There were 112 deaths associated with the construction of the dam. The first was J. G. Tierney, a surveyor who drowned on December 20, 1922, while looking for an ideal spot for the dam. The last death on the project's official fatality list occurred on December 20, 1935 when an "electrician's helper", Patrick Tierney, the son of J.G. Tierney, fell from an intake tower. Included in the fatality list are three workers, one in 1932 and two in 1933, who committed suicide onsite. Ninety-six of the deaths occurred during construction at the site. Of the 112 fatalities, 91 were Six Companies employees, three were BOR employees, and one was a visitor to the site, with the remainder employees of various contractors not part of Six Companies.

Not included in the official number of fatalities were deaths that were recorded as pneumonia. Workers alleged that this diagnosis was a cover for death from carbon monoxide poisoning (brought on by the use of gasoline-fueled vehicles in the diversion tunnels), and a classification used by Six Companies to avoid paying compensation claims. The site's diversion tunnels frequently reached 140 °F (60 °C), enveloped in thick plumes of vehicle exhaust gases. A total of 42 workers were recorded as having died from pneumonia; none were listed as having died from carbon monoxide poisoning. No deaths of non-workers from pneumonia were recorded in Boulder City during the construction period.

Architectural Style

Hansen's bas-relief on the Nevada elevator

The initial plans for the facade of the dam, the power plant, the outlet tunnels and ornaments clashed with the modern look of an arch dam. The Bureau of Reclamation, more concerned with the dam's functionality, adorned it with a Gothic-inspired balustrade and eagle statues. This initial design was criticized by many as being too plain and unremarkable for a project of such immense scale, so Los Angeles-based architect Gordon B. Kaufmann, then the supervising architect to the Bureau of Reclamation, was brought in to redesign the exteriors. Kaufmann greatly streamlined the design, and applied an elegant Art Deco style to the entire project. He designed sculptured turrets rising seamlessly from the dam face and clock faces on the intake towers set for the time in Nevada and Arizona — both states are in different time zones, but since Arizona does not observe Daylight Saving Time, the clocks display the same time for more than half the year.

At Kaufmann's request, Denver artist Allen Tupper True was hired to handle the design and decoration of the walls and floors of the new dam. True's design scheme incorporated motifs of the Navajo and Pueblo tribes of the region. Although some initially were opposed to these designs, True was given the go-ahead and was officially appointed consulting artist. With the assistance of the National Laboratory of Anthropology, True researched authentic decorative motifs from Indian sand paintings, textiles, baskets and ceramics. The images and colors are based on Native

American visions of rain, lightning, water, clouds, and local animals — lizards, serpents, birds — and on the Southwestern landscape of stepped mesas. In these works, which are integrated into the walkways and interior halls of the dam, True also reflected on the machinery of the operation, making the symbolic patterns appear both ancient and modern.

Tile floor designed by Allen Tupper True

With the agreement of Kaufmann and the engineers, True also devised an innovative color-coding for the pipes and machinery, which was implemented throughout all BOR projects. True's consulting artist job lasted through 1942; it was extended so he could complete design work for the Parker, Shasta and Grand Coulee dams and power plants. True's work on the Hoover Dam was humorously referred to in a poem published in *The New Yorker*, part of which read, "lose the spark, and justify the dream; but also worthy of remark will be the color scheme".

Complementing Kaufmann and True's work, the Norwegian-born, naturalized American sculptor Oskar J.W. Hansen designed many of the sculptures on and around the dam. His works include the monument of dedication plaza, a plaque to memorialize the workers killed and the bas-reliefs on the elevator towers. In his words, Hansen wanted his work to express "the immutable calm of intellectual resolution, and the enormous power of trained physical strength, equally enthroned in placid triumph of scientific accomplishment", because "[t]he building of Hoover Dam belongs to the sagas of the daring." Hansen's dedication plaza, on the Nevada abutment, contains a sculpture of two winged figures flanking a flagpole.

Hoover Dam memorial star map floor, center area

Surrounding the base of the monument is a terrazzo floor embedded with a "star map". The map depicts the Northern Hemisphere sky at the moment of President Roosevelt's dedication of the dam. This is intended to help future astronomers, if necessary, calculate the exact date of dedication. The 30-foot-high (9.1 m) bronze figures, dubbed "Winged Figures of the Republic", were each formed in a continuous pour. To put such large bronzes into place without marring the highly polished bronze surface, they were placed on ice and guided into position as the ice melted. Hansen's bas-relief on the Nevada elevator tower depicts the benefits of the dam: flood control, navigation, irrigation, water storage, and power. The bas-relief on the Arizona elevator depicts, in his words, "the visages of those Indian tribes who have inhabited mountains and plains from ages distant."

Operation

Power Plant and Water Demands

Hoover Dam releasing water from the jet-flow gates in 1998

Excavation for the powerhouse was carried out simultaneously with the excavation for the dam foundation and abutments. A U-shaped structure located at the downstream toe of the dam, its excavation was completed in late 1933 with the first concrete placed in November 1933. Filling of Lake Mead began February 1, 1935, even before the last of the concrete was poured that May. The powerhouse was one of the projects uncompleted at the time of the formal dedication on September 30, 1935—a crew of 500 men remained to finish it and other structures. To make the powerhouse roof bombproof, it was constructed of layers of concrete, rock, and steel with a total thickness of about 3.5 feet (1.1 m), topped with layers of sand and tar.

In the latter half of 1936, water levels in Lake Mead were high enough to permit power generation, and the first three Allis Chalmers built Francis turbine-generators, all on the Nevada side, began operating. In March 1937, one more Nevada generator went online and the first Arizona generator by August. By September 1939, four more generators were operating, and the dam's power plant became the largest hydroelectricity facility in the world. The final generator was not placed in service until 1961, bringing the maximum generating capacity to 1,345 megawatts at the time. Original plans called for 16 large generators, eight on each side of the river, but two smaller generators were installed instead of one large one on the Arizona side for a total of 17. The smaller generators were used to serve smaller communities at a time when the output of each generator was dedicated to a single municipality, before the dam's total power output was placed on the grid and made arbitrarily distributable. The present contracts for the sale of electricity expire in 2017.

Before water from Lake Mead reaches the turbines, it enters the intake towers and then four gradually narrowing penstocks which funnel the water down towards the powerhouse. The intakes provide a maximum hydraulic head (water pressure) of 590 ft (180 m) as the water reaches a speed of about 85 mph (140 km/h). The entire flow of the Colorado River passes through the turbines. The spillways and outlet works (jet-flow gates) are rarely used. The jet-flow gates, located in concrete structures 180 feet (55 m) above the river, and also at the outlets of the inner diversion tunnels at river level, may be used to divert water around the dam in emergency or flood conditions, but have never done so, and in practice are only used to drain water from the penstocks for maintenance. Following an uprating project from 1986 to 1993, the total gross power rating for the plant, including two 2.4 megawatt Pelton turbine-generators that power Hoover Dam's own operations is a maximum capacity of 2080 megawatts. The annual generation of Hoover Dam varies. The maximum net generation was 10.348 TWh in 1984, and the minimum since 1940 was 2.648 TWh in 1956. The average power generated was 4.2 TWh/year for 1947-2008. In 2015, the dam generated 3.6 TWh.

The amount of electricity generated by Hoover Dam has been decreasing along with the falling water level in Lake Mead due to the prolonged drought in the 2010s and high demand for the Colorado River's water. Lake Mead fell to a new record low elevation of 1,071.61 feet (326.63 m) on July 1, 2016 before beginning to rebound slowly. Under its original design, the dam will no longer be able to generate power once the water level falls below 1,050 feet (320 m), which could occur as early as 2017. To lower the minimum power pool elevation from 1,050 to 950 feet (320 to 290 m), five wide-head turbines, designed to work efficiently with less flow, are being installed and will be fully online by 2017. Due to the low water levels, by 2014 it was providing power only during periods of peak demand.

Control of water was the primary concern in the building of the dam. Power generation has allowed the dam project to be self-sustaining: proceeds from the sale of power repaid the 50-year construction loan, and those revenues also finance the multimillion-dollar yearly maintenance budget. Power is generated in step with and only with the release of water in response to downstream water demands.

Lake Mead and downstream releases from the dam also provide water for both municipal and irrigation uses. Water released from the Hoover Dam eventually reaches several canals. The Colorado River Aqueduct and Central Arizona Project branch off Lake Havasu while the All-American Canal is supplied by the Imperial Dam. In total, water from the Lake Mead serves 18 million people in Arizona, Nevada and California and supplies the irrigation of over 1,000,000 acres (400,000 ha) of land.

Power Distribution

Electricity from the dam's powerhouse was originally sold pursuant to a fifty-year contract, authorized by Congress in 1934, which ran from 1937 to 1987. In 1984, Congress passed a new statute which set power allocations from the dam from 1987 to 2017. The powerhouse was run under the original authorization by the Los Angeles Department of Water and Power and Southern California Edison; in 1987, the Bureau of Reclamation assumed control. In 2011, Congress enacted legislation extending the current contracts until 2067, after setting aside 5% of Hoover Dam's power for sale to Native American tribes, electric cooperatives, and other entities. The new arrangement will begin in 2017. The Bureau of Reclamation reports that the energy generated is allocated as follows:

Area	Percentage
Metropolitan Water District of Southern California	28.53%
State of Nevada	23.37%
State of Arizona	18.95%
Los Angeles, California	15.42%
Southern California Edison	5.54%
Boulder City, Nevada	1.77%
Glendale, California	1.59%
Pasadena, California	1.36%
Anaheim, California	1.15%
Riverside, California	0.86%
Vernon, California	0.62%
Burbank, California	0.59%
Azusa, California	0.11%
Colton, California	0.09%
Banning, California	0.05%

Tourists gather around one of the generators in the Nevada wing of the powerhouse to hear its operation explained, September 1940

A worker stands by the 30 ft (9.1 m) diameter Nevada penstock before its junction with another penstock that delivers water to a turbine

Spillways

Water enters the Arizona spillway (left) during the 1983 floods. Lake Mead water level was 1,225.6 ft (373.6 m)

The dam is protected against over-topping by two spillways. The spillway entrances are located behind each dam abutment, running roughly parallel to the canyon walls. The spillway entrance arrangement forms a classic side-flow weir with each spillway containing four 100-foot-long (30 m) and 16-foot-wide (4.9 m) steel-drum gates. Each gate weighs 5,000,000 pounds (2,300,000 kg) and can be operated manually or automatically. Gates are raised and lowered depending on water levels in the reservoir and flood conditions. The gates could not entirely prevent water from entering the spillways but could maintain an extra 16 ft (4.9 m) of lake level. Water flowing over the spillways falls dramatically into 600-foot-long (180 m), 50-foot-wide (15 m) spillway tunnels before connecting to the outer diversion tunnels, and reentering the main river channel below the dam. This complex spillway entrance arrangement combined with the approximate 700-foot (210 m) elevation drop from the top of the reservoir to the river below was a difficult engineering problem and posed numerous design challenges. Each spillway's capacity of 200,000 cu ft/s (5,700 m³/s) was empirically verified in post-construction tests in 1941.

The large spillway tunnels have only been used twice, for testing in 1941 and because of flooding in 1983. During both times, when inspecting the tunnels after the spillways were used, engineers found major damage to the concrete linings and underlying rock. The 1941 damage was attributed to a slight misalignment of the tunnel invert (or base), which caused cavitation, a phenomenon in fast-flowing liquids in which vapor bubbles collapse with explosive force. In response to this finding, the tunnels were patched with special heavy-duty concrete and the surface of the concrete was polished mirror-smooth. The spillways were modified in 1947 by adding flip buckets, which both slow the water and decrease the spillway's effective capacity, in an attempt to eliminate conditions thought to have contributed to the 1941 damage. The 1983 damage, also due to cavitation, led to the installation of aerators in the spillways. Tests at Grand Coulee Dam showed that the technique worked, in principle.

Roadway and Tourism

The bypass in front of the dam

There are two lanes for automobile traffic across the top of the dam, which formerly served as the Colorado River crossing for U.S. Route 93. In the wake of the September 11, 2001 terrorist attacks, authorities expressed security concerns and the Hoover Dam Bypass project was expedited. Pending the completion of the bypass, restricted traffic was permitted over Hoover Dam. Some types of vehicles were inspected prior to crossing the dam while semi-trailer trucks, buses carrying luggage, and enclosed-box trucks over 40 ft (12 m) long were not allowed on the dam at all, and were diverted to U.S. Route 95 or Nevada State Routes 163/68. The four-lane Hoover Dam Bypass opened on October 19, 2010. It includes a composite steel and concrete arch bridge, the Mike O'Callaghan–Pat Tillman Memorial Bridge, 1,500 ft (460 m) downstream from the dam. With the opening of the bypass, through traffic is no longer allowed across Hoover Dam, dam visitors are allowed to use the existing roadway to approach from the Nevada side and cross to parking lots and other facilities on the Arizona side.

Hoover Dam opened for tours in 1937 after its completion, but following Japan's attack on Pearl Harbor on December 7, 1941, it was closed to the public when the United States entered World War II, during which only authorized traffic, in convoys, was permitted. After the war, it reopened September 2, 1945, and by 1953, annual attendance had risen to 448,081. The dam closed on November 25, 1963 and March 31, 1969, days of mourning in remembrance of Presidents Kennedy and Eisenhower. In 1995, a new visitors' center was built, and the following year, visits exceeded one million for the first time. The dam closed again to the public on September 11, 2001; modified tours were resumed in December and a new "Discovery Tour" was added the following year. Today, nearly a million people per year take the tours of the dam offered by the Bureau of Reclamation. Increased security concerns by the government have led to most of the interior structure being inaccessible to tourists. As a result, few of True's decorations can now be seen by visitors.

Environmental Impact

View upstream from Hoover Dam, Sept. 2009. Water elevation on this date was 1093.77 ft.

The changes in water flow and use caused by Hoover Dam's construction and operation have had a large impact on the Colorado River Delta. The construction of the dam has been implicated in causing the decline of this estuarine ecosystem. For six years after the construction of the dam, while Lake Mead filled, virtually no water reached the mouth of the river. The delta's estuary, which once had a freshwater-saltwater mixing zone stretching 40 miles (64 km) south of the river's mouth, was turned into an inverse estuary where the level of salinity was higher close to the river's mouth.

The Colorado River had experienced natural flooding before the construction of the Hoover Dam. The dam eliminated the natural flooding, which threatened many species adapted to the flooding, including both plants and animals. The construction of the dam devastated the populations of native fish in the river downstream from the dam. Four species of fish native to the Colorado River, the Bonytail chub, Colorado pikeminnow, Humpback chub, and Razorback sucker, are listed as endangered.

Naming Controversy

During the years of lobbying leading up to the passage of legislation authorizing the dam in 1928, the press generally referred to the dam as "Boulder Dam" or as "Boulder Canyon Dam", even though the proposed site had shifted to Black Canyon. The Boulder Canyon Project Act of 1928 (BCPA) never mentioned a proposed name or title for the dam. The BCPA merely allows the government to "construct, operate, and maintain a dam and incidental works in the main stream of the Colorado River at Black Canyon or Boulder Canyon".

When Secretary Wilbur spoke at the ceremony starting the building of the railway between Las Vegas and the dam site on September 17, 1930, he named the dam "Hoover Dam", citing a tradition of naming dams after Presidents, though none had been so honored during their terms of office. Wilbur justified his choice on the ground that Hoover was "the great engineer whose vision and persistence ... has done so much to make [the dam] possible". One writer complained in response that "the Great Engineer had quickly drained, ditched, and dammed the country."

After Hoover's election defeat in 1932 and the accession of the Roosevelt administration, Secretary Ickes ordered on May 13, 1933, that the dam be referred to as "Boulder Dam". Ickes stated that Wilbur had been imprudent in naming the dam after a sitting president, that Congress had never ratified his choice, and that it had long been referred to as Boulder Dam. Unknown to the general public, Attorney General Homer Cummings informed Ickes that Congress had indeed used the name "Hoover Dam" in five different bills appropriating money for construction of the dam. The official status this conferred to the name "Hoover Dam" had been noted on the floor of the House of Representatives by Congressman Edward T. Taylor of Colorado on December 12, 1930, but was likewise ignored by Ickes.

When Ickes spoke at the dedication ceremony on September 30, 1935, he was determined, as he recorded in his diary, "to try to nail down for good and all the name Boulder Dam." At one point in the speech, he spoke the words "Boulder Dam" five times within thirty seconds. Further, he suggested that if the dam were to be named after any one person, it should be for California Senator Hiram Johnson, a lead sponsor of the authorizing legislation. Roosevelt also referred to the dam as Boulder Dam, and the Republican-leaning *Los Angeles Times*, which at the time of Ickes' name change had run an editorial cartoon showing Ickes ineffectively chipping away at an enormous sign "HOOVER DAM," reran it showing Roosevelt reinforcing Ickes, but having no greater success.

In the following years, the name "Boulder Dam" failed to fully take hold, with many Americans using both names interchangeably and mapmakers divided as to which name should be printed. Memories of the Great Depression faded, and Hoover to some extent rehabilitated himself through good works during and after World War II. In 1947, a bill passed both Houses of Congress unanimously restoring the name "Hoover Dam." Ickes, who was by then a private citizen, opposed the change, stating, "I didn't know Hoover was that small a man to take credit for something he had nothing to do with."

Hoover Dam 2011 panoramic view from the Arizona side showing the penstock towers, the Nevada-side spillway entrance and the Mike O'Callaghan – Pat Tillman Memorial Bridge, also known as the Hoover Dam Bypass

References

- Rodney White (1 January 2001). Evacuation of Sediments from Reservoirs. Thomas Telford Publishing. pp. 163–169. ISBN 978-0727729538. Retrieved 4 August 2013.

- Burns, William C. G. (2001). The World's Water, 2002–2003: The Biennial Report on Freshwater Resources. Washington DC: Island Press. p. 139. ISBN 978-1-55963-949-1.

- "The world's largest "L boat lift" Three Gorges Dam successfully tested" (in Chinese). Chutianjinbao News. January 14, 2016. Retrieved February 15, 2016.

- "Phase I Field Trial of Ship Lift at Three Gorges Dam Successfully Ends". CHINA THREE GORGES PROJECT. 2016-08-14. Retrieved 2016-08-14.

- Teixeira, Cristiano (5 April 2016), Corredor Ecológico de Santa Maria, Paraná - Brasil (PDF) (in Portuguese), Asunción: ITAIPU Binacional/MI, p. 3, retrieved 2016-11-04

- "World Bank approves $390m loan for Tarbela fifth extension - The Express Tribune". 2016-09-21. Retrieved 2016-09-22.

- Reporter, The Newspaper's Staff (7 March 2015). "Another $51m to expedite Tarbela project completion". Retrieved 30 August 2016.

- "Tackling energy crisis: Nawaz inaugurates Tarbela-IV project". The Express Tribune. 26 February 2014. Retrieved 23 March 2016.

- "Tarbela 4th, 5th extension projects to add 2820 MW to system: Abid". The Nation (Pakistan). 15 February 2016. Retrieved 23 March 2016.

- Harris, Michael (2 October 2015). "Pakistan's 1,300-MW Tarbela 5th Extension hydropower project receives next-to-last approval". Hydroworld. Retrieved 23 March 2016.

- "Pakistan extends consultant call for 1,300-MW Tarbela Dam 5th Extension". Hydroworld. 19 August 2014. Retrieved 23 March 2016.

- "WAPDA to submit $795.8m PC-1 of Tarbela 5th extension project for approval". The Daily Times. 26 March 2015. Retrieved 23 March 2016.

- "OPERATION PLAN FOR COLORADO RIVER SYSTEM RESERVOIRS" (PDF). Bureau of Reclamation. December 2015. Retrieved 7 June 2016.

- Kuckro, Rod (June 30, 2014). "Receding Lake Mead poses challenges to Hoover Dam's power output". E&E Publishing, LLC. Retrieved June 7, 2016.

Hydroelectricity: An Overview

Hydroelectricity is the electricity that is produced from hydropower. It is produced in the majority of the countries. Run-of-the-river hydroelectricity, small hydro, micro hydro, pico hydro, underground power station and energy storage are the aspects that have been explicated in the chapter.

Hydroelectricity

The Three Gorges Dam in Central China is the world's largest power producing facility of any kind.

Hydroelectricity is electricity produced from hydropower. In 2015 hydropower generated 16.6% of the world's total electricity and 70% of all renewable electricity, and was expected to increase about 3.1% each year for the next 25 years.

Hydropower is produced in 150 countries, with the Asia-Pacific region generating 33 percent of global hydropower in 2013. China is the largest hydroelectricity producer, with 920 TWh of production in 2013, representing 16.9 percent of domestic electricity use.

The cost of hydroelectricity is relatively low, making it a competitive source of renewable electricity. The hydro station consumes no water, unlike coal or gas plants. The average cost of electricity from a hydro station larger than 10 megawatts is 3 to 5 U.S. cents per kilowatt-hour. With a dam and reservoir it is also a flexible source of electricity since the amount produced by the station can be changed up or down very quickly to adapt to changing energy demands. Once a hydroelectric complex is constructed, the project produces no direct waste, and has a considerably lower output level of greenhouse gases than fossil fuel powered energy plants.

History

Hydropower has been used since ancient times to grind flour and perform other tasks. In the mid-1770s, French engineer Bernard Forest de Bélidor published *Architecture Hydraulique* which described vertical- and horizontal-axis hydraulic machines. By the late 19th century, the electrical generator was developed and could now be coupled with hydraulics. The growing demand for the Industrial Revolution would drive development as well. In 1878 the world's first hydroelectric power scheme was developed at Cragside in Northumberland, England by William George Armstrong. It was used to power a single arc lamp in his art gallery. The old Schoelkopf Power Station No. 1 near Niagara Falls in the U.S. side began to produce electricity in 1881. The first Edison hydroelectric power station, the Vulcan Street Plant, began operating September 30, 1882, in Appleton, Wisconsin, with an output of about 12.5 kilowatts. By 1886 there were 45 hydroelectric power stations in the U.S. and Canada. By 1889 there were 200 in the U.S. alone.

Museum Hydroelectric power plant "Under the Town" in Serbia, built in 1900.

At the beginning of the 20th century, many small hydroelectric power stations were being constructed by commercial companies in mountains near metropolitan areas. Grenoble, France held the International Exhibition of Hydropower and Tourism with over one million visitors. By 1920 as 40% of the power produced in the United States was hydroelectric, the Federal Power Act was enacted into law. The Act created the Federal Power Commission to regulate hydroelectric power stations on federal land and water. As the power stations became larger, their associated dams developed additional purposes to include flood control, irrigation and navigation. Federal funding became necessary for large-scale development and federally owned corporations, such as the Tennessee Valley Authority (1933) and the Bonneville Power Administration (1937) were created. Additionally, the Bureau of Reclamation which had begun a series of western U.S. irrigation projects in the early 20th century was now constructing large hydroelectric projects such as the 1928 Hoover Dam. The U.S. Army Corps of Engineers was also involved in hydroelectric development, completing the Bonneville Dam in 1937 and being recognized by the Flood Control Act of 1936 as the premier federal flood control agency.

Hydroelectric power stations continued to become larger throughout the 20th century. Hydropower was referred to as *white coal* for its power and plenty. Hoover Dam's initial 1,345 MW power station was the world's largest hydroelectric power station in 1936; it was eclipsed by the 6809 MW Grand Coulee Dam in 1942. The Itaipu Dam opened in 1984 in South America as the largest, producing 14,000 MW but was surpassed in 2008 by the Three Gorges Dam in China at 22,500 MW. Hydroelectricity would eventually supply some countries, including Norway, Democratic Republic of the Congo, Paraguay and Brazil, with over 85% of their electricity. The United States currently has over 2,000 hydroelectric power stations that supply 6.4% of its total electrical production output, which is 49% of its renewable electricity.

Generating Methods

Turbine row at El Nihuil II Power Station in Mendoza, Argentina

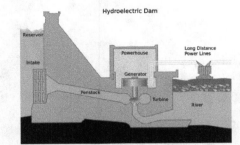

Cross section of a conventional hydroelectric dam.

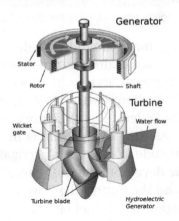

A typical turbine and generator

Conventional (Dams)

Most hydroelectric power comes from the potential energy of dammed water driving a water turbine and generator. The power extracted from the water depends on the volume and on the difference in height between the source and the water's outflow. This height difference is called the head. A large pipe (the "penstock") delivers water from the reservoir to the turbine.

Pumped-storage

This method produces electricity to supply high peak demands by moving water between reservoirs at different elevations. At times of low electrical demand, the excess generation capacity is used to pump water into the higher reservoir. When the demand becomes greater, water is released back into the lower reservoir through a turbine. Pumped-storage schemes currently provide the most commercially important means of large-scale grid energy storage and improve the daily capacity factor of the generation system. Pumped storage is not an energy source, and appears as a negative number in listings.

Run-of-the-river

Run-of-the-river hydroelectric stations are those with small or no reservoir capacity, so that only the water coming from upstream is available for generation at that moment, and any oversupply must pass unused. A constant supply of water from a lake or existing reservoir upstream is a significant advantage in choosing sites for run-of-the-river. In the United States, run of the river hydropower could potentially provide 60,000 megawatts (80,000,000 hp) (about 13.7% of total use in 2011 if continuously available).

Tide

A tidal power station makes use of the daily rise and fall of ocean water due to tides; such sources are highly predictable, and if conditions permit construction of reservoirs, can also be dispatchable to generate power during high demand periods. Less common types of hydro schemes use water's kinetic energy or undammed sources such as undershot water wheels. Tidal power is viable in a relatively small number of locations around the world. In Great Britain, there are eight sites that could be developed, which have the potential to generate 20% of the electricity used in 2012.

Sizes, Types and Capacities of Hydroelectric Facilities

Large Facilities

Large-scale hydroelectric power stations are more commonly seen as the largest power producing facilities in the world, with some hydroelectric facilities capable of generating more than double the installed capacities of the current largest nuclear power stations.

Although no official definition exists for the capacity range of large hydroelectric power stations, facilities from over a few hundred megawatts are generally considered large hydroelectric facilities.

Currently, only four facilities over 10 GW (10,000 MW) are in operation worldwide, see table below.

Rank	Station	Country	Location	Capacity (MW)
1.	Three Gorges Dam	China	30°49′15″N 111°00′08″E30.82083°N 111.00222°E	22,500
2.	Itaipu Dam	Brazil Paraguay	25°24′31″S 54°35′21″W25.40861°S 54.58917°W	14,000
3.	Xiluodu Dam	China	28°15′35″N 103°38′58″E28.25972°N 103.64944°E	13,860
4.	Guri Dam	Venezuela	07°45′59″N 62°59′57″W7.76639°N 62.99917°W	10,200

Panoramic view of the Itaipu Dam, with the spillways (closed at the time of the photo) on the left. In 1994, the American Society of Civil Engineers elected the Itaipu Dam as one of the seven modern Wonders of the World.

Small

Small hydro is the development of hydroelectric power on a scale serving a small community or industrial plant. The definition of a small hydro project varies but a generating capacity of up to 10 megawatts (MW) is generally accepted as the upper limit of what can be termed small hydro. This may be stretched to 25 MW and 30 MW in Canada and the United States. Small-scale hydro-electricity production grew by 28% during 2008 from 2005, raising the total world small-hydro capacity to 85 GW. Over 70% of this was in China (65 GW), followed by Japan (3.5 GW), the United States (3 GW), and India (2 GW).

A micro-hydro facility in Vietnam

Pico hydroelectricity in Mondulkiri, Cambodia

Small hydro stations may be connected to conventional electrical distribution networks as a source of low-cost renewable energy. Alternatively, small hydro projects may be built in isolated areas that would be uneconomic to serve from a network, or in areas where there is no national electrical distribution network. Since small hydro projects usually have minimal reservoirs and civil construction work, they are seen as having a relatively low environmental impact compared to large hydro. This decreased environmental impact depends strongly on the balance between stream flow and power production.

Micro

Micro hydro is a term used for hydroelectric power installations that typically produce up to 100 kW of power. These installations can provide power to an isolated home or small community, or are sometimes connected to electric power networks. There are many of these installations around the world, particularly in developing nations as they can provide an economical source of energy without purchase of fuel. Micro hydro systems complement photovoltaic solar energy systems because in many areas, water flow, and thus available hydro power, is highest in the winter when solar energy is at a minimum.

Pico

Pico hydro is a term used for hydroelectric power generation of under 5 kW. It is useful in small, remote communities that require only a small amount of electricity. For example, to power one or two fluorescent light bulbs and a TV or radio for a few homes. Even smaller turbines of 200-300W may power a single home in a developing country with a drop of only 1 m (3 ft). A Pico-hydro setup is typically run-of-the-river, meaning that dams are not used, but rather pipes divert some of the flow, drop this down a gradient, and through the turbine before returning it to the stream.

Underground

An underground power station is generally used at large facilities and makes use of a large natural height difference between two waterways, such as a waterfall or mountain lake. An underground tunnel is constructed to take water from the high reservoir to the generating hall built in an underground cavern near the lowest point of the water tunnel and a horizontal tailrace taking water away to the lower outlet waterway.

Measurement of the tailrace and forebay rates at the Limestone Generating Station in Manitoba, Canada.

Calculating Available Power

A simple formula for approximating electric power production at a hydroelectric station is: $P = \rho h r g k$, where

- P is Power in watts,

- ρ is the density of water (~1000 kg/m³),

- h is height in meters,

- *r* is flow rate in cubic meters per second,

- *g* is acceleration due to gravity of 9.8 m/s²,

-    is a coefficient of efficiency ranging from 0 to 1. Efficiency is often higher (that is, closer to 1) with larger and more modern turbines.

Annual electric energy production depends on the available water supply. In some installations, the water flow rate can vary by a factor of 10:1 over the course of a year.

Properties

Advantages

The Ffestiniog Power Station can generate 360 MW of electricity within 60 seconds of the demand arising.

Flexibility

Hydropower is a flexible source of electricity since stations can be ramped up and down very quickly to adapt to changing energy demands. Hydro turbines have a start-up time of the order of a few minutes. It takes around 60 to 90 seconds to bring a unit from cold start-up to full load; this is much shorter than for gas turbines or steam plants. Power generation can also be decreased quickly when there is a surplus power generation. Hence the limited capacity of hydropower units is not generally used to produce base power except for vacating the flood pool or meeting downstream needs. Instead, it serves as backup for non-hydro generators.

Low Cost/High Value Power

The major advantage of conventional hydroelectric dams with reservoirs is their ability to store water at low cost for dispatch later as high value clean electricity. The average cost of electricity from a hydro station larger than 10 megawatts is 3 to 5 U.S. cents per kilowatt-hour. When used as peak power to meet demand, hydroelectricity has a higher value than base power and a much higher value compared to intermittent energy sources.

Hydroelectric stations have long economic lives, with some plants still in service after 50–100 years. Operating labor cost is also usually low, as plants are automated and have few personnel on site during normal operation.

Where a dam serves multiple purposes, a hydroelectric station may be added with relatively low construction cost, providing a useful revenue stream to offset the costs of dam operation. It has been calculated that the sale of electricity from the Three Gorges Dam will cover the construction costs after 5 to 8 years of full generation. Additionally, some data shows that in most countries large hydropower dams will be too costly and take too long to build to deliver a positive risk adjusted return, unless appropriate risk management measures are put in place.

Suitability for Industrial Applications

While many hydroelectric projects supply public electricity networks, some are created to serve specific industrial enterprises. Dedicated hydroelectric projects are often built to provide the substantial amounts of electricity needed for aluminium electrolytic plants, for example. The Grand Coulee Dam switched to support Alcoa aluminium in Bellingham, Washington, United States for American World War II airplanes before it was allowed to provide irrigation and power to citizens (in addition to aluminium power) after the war. In Suriname, the Brokopondo Reservoir was constructed to provide electricity for the Alcoa aluminium industry. New Zealand's Manapouri Power Station was constructed to supply electricity to the aluminium smelter at Tiwai Point.

Reduced CO_2 Emissions

Since hydroelectric dams do not burn fossil fuels, they do not directly produce carbon dioxide. While some carbon dioxide is produced during manufacture and construction of the project, this is a tiny fraction of the operating emissions of equivalent fossil-fuel electricity generation. One measurement of greenhouse gas related and other externality comparison between energy sources can be found in the ExternE project by the Paul Scherrer Institute and the University of Stuttgart which was funded by the European Commission. According to that study, hydroelectricity produces the least amount of greenhouse gases and externality of any energy source. Coming in second place was wind, third was nuclear energy, and fourth was solar photovoltaic. The low greenhouse gas impact of hydroelectricity is found especially in temperate climates. The above study was for local energy in Europe; presumably similar conditions prevail in North America and Northern Asia, which all see a regular, natural freeze/thaw cycle (with associated seasonal plant decay and regrowth). Greater greenhouse gas emission impacts are found in the tropical regions because the reservoirs of power stations in tropical regions produce a larger amount of methane than those in temperate areas.

Other uses of the Reservoir

Reservoirs created by hydroelectric schemes often provide facilities for water sports, and become tourist attractions themselves. In some countries, aquaculture in reservoirs is common. Multi-use dams installed for irrigation support agriculture with a relatively constant water supply. Large hydro dams can control floods, which would otherwise affect people living downstream of the project.

Disadvantages

Ecosystem Damage and Loss of Land

Large reservoirs associated with traditional hydroelectric power stations result in submersion of extensive areas upstream of the dams, sometimes destroying biologically rich and productive lowland and riverine valley forests, marshland and grasslands. Damming interrupts the flow of rivers and can harm local ecosystems, and building large dams and reservoirs often involves displacing people and wildlife. The loss of land is often exacerbated by habitat fragmentation of surrounding areas caused by the reservoir.

Hydroelectric power stations that use dams would submerge large areas of land due to the requirement of a reservoir. Merowe Dam in Sudan.

Hydroelectric projects can be disruptive to surrounding aquatic ecosystems both upstream and downstream of the plant site. Generation of hydroelectric power changes the downstream river environment. Water exiting a turbine usually contains very little suspended sediment, which can lead to scouring of river beds and loss of riverbanks. Since turbine gates are often opened intermittently, rapid or even daily fluctuations in river flow are observed.

Siltation and Flow Shortage

When water flows it has the ability to transport particles heavier than itself downstream. This has a negative effect on dams and subsequently their power stations, particularly those on rivers or within catchment areas with high siltation. Siltation can fill a reservoir and reduce its capacity to control floods along with causing additional horizontal pressure on the upstream portion of the dam. Eventually, some reservoirs can become full of sediment and useless or over-top during a flood and fail.

Changes in the amount of river flow will correlate with the amount of energy produced by a dam. Lower river flows will reduce the amount of live storage in a reservoir therefore reducing the amount of water that can be used for hydroelectricity. The result of diminished river flow can be power shortages in areas that depend heavily on hydroelectric power. The risk of flow shortage may increase as a result of climate change. One study from the Colorado River in the United States suggest that modest climate changes, such as an increase in temperature in 2 degree Celsius resulting in a 10% decline in precipitation, might reduce river run-off by up to 40%. Brazil in particular is vulnerable due to its heavy reliance on hydroelectricity, as increasing temperatures, lower water flow and alterations in the rainfall regime, could reduce total energy production by 7% annually by the end of the century.

Methane Emissions (From Reservoirs)

Lower positive impacts are found in the tropical regions, as it has been noted that the reservoirs of power plants in tropical regions produce substantial amounts of methane. This is due to plant material in flooded areas decaying in an anaerobic environment, and forming methane, a greenhouse gas. According to the World Commission on Dams report, where the reservoir is large compared to the generating capacity (less than 100 watts per square metre of surface area) and no clearing of the forests in the area was undertaken prior to impoundment of the reservoir, greenhouse gas emissions from the reservoir may be higher than those of a conventional oil-fired thermal generation plant.

The Hoover Dam in the United States is a large conventional dammed-hydro facility, with an installed capacity of 2,080 MW.

In boreal reservoirs of Canada and Northern Europe, however, greenhouse gas emissions are typically only 2% to 8% of any kind of conventional fossil-fuel thermal generation. A new class of underwater logging operation that targets drowned forests can mitigate the effect of forest decay.

Relocation

Another disadvantage of hydroelectric dams is the need to relocate the people living where the reservoirs are planned. In 2000, the World Commission on Dams estimated that dams had physically displaced 40-80 million people worldwide.

Failure Risks

Because large conventional dammed-hydro facilities hold back large volumes of water, a failure due to poor construction, natural disasters or sabotage can be catastrophic to downriver settlements and infrastructure.

During Typhoon Nina in 1975 Banqiao Dam failed in Southern China when more than a year's worth of rain fell within 24 hours. The resulting flood resulted in the deaths of 26,000 people, and another 145,000 from epidemics. Millions were left homeless.

The creation of a dam in a geologically inappropriate location may cause disasters such as 1963 disaster at Vajont Dam in Italy, where almost 2,000 people died.

The Malpasset Dam failure in Fréjus on the French Riviera (Côte d'Azur), southern France, collapsed on December 2, 1959, killing 423 people in the resulting flood.

Smaller dams and micro hydro facilities create less risk, but can form continuing hazards even after being decommissioned. For example, the small Kelly Barnes Dam failed in 1977, causing 39 deaths with the Toccoa Flood, twenty years after its power station was decommissioned the earthen embankment dam failed.

Comparison with other Methods of Power Generation

Hydroelectricity eliminates the flue gas emissions from fossil fuel combustion, including pollutants such as sulfur dioxide, nitric oxide, carbon monoxide, dust, and mercury in the coal. Hydroelectricity also avoids the hazards of coal mining and the indirect health effects of coal emissions. Compared to nuclear power, hydroelectricity construction requires altering large areas of the environment while a nuclear power station has a small footprint, and hydro-powerstation failures have caused tens of thousands of more deaths than any nuclear station failure. The creation of Garrison Dam, for example, required Native American land to create Lake Sakakawea, which has a shoreline of 1,320 miles, and caused the inhabitants to sell 94% of their arable land for $7.5 million in 1949.

Compared to wind farms, hydroelectricity power stations have a more predictable load factor. If the project has a storage reservoir, it can generate power when needed. Hydroelectric stations can be easily regulated to follow variations in power demand.

World Hydroelectric Capacity

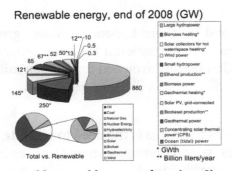

World renewable energy share (2008)

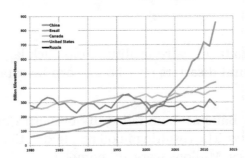

Trends in the top five hydroelectricity-producing countries

The ranking of hydro-electric capacity is either by actual annual energy production or by installed capacity power rating. In 2015 hydropower generated 16.6% of the worlds total electricity and 70% of all renewable electricity. Hydropower is produced in 150 countries, with the Asia-Pacific region generated 32 percent of global hydropower in 2010. China is the largest hydroelectricity producer, with 721 terawatt-hours of production in 2010, representing around 17 percent of domestic electricity use. Brazil, Canada, New Zealand, Norway, Paraguay, Austria, Switzerland, and Venezuela have a majority of the internal electric energy production from hydroelectric power. Paraguay produces 100% of its electricity from hydroelectric dams, and exports 90% of its production to Brazil and to Argentina. Norway produces 98–99% of its electricity from hydroelectric sources.

A hydro-electric station rarely operates at its full power rating over a full year; the ratio between annual average power and installed capacity rating is the capacity factor. The installed capacity is the sum of all generator nameplate power ratings.

Ten of the largest hydroelectric producers as at 2014.				
Country	Annual hydroelectric production (TWh)	Installed capacity (GW)	Capacity factor	% of total production
China	1064	311	0.37	18.7%
Canada	383	76	0.59	58.3%
Brazil	373	89	0.56	63.2%
United States	282	102	0.42	6.5%
Russia	177	51	0.42	16.7%
India	132	40	0.43	10.2%
Norway	129	31	0.49	96.0%
Japan	87	50	0.37	8.4%
Venezuela	87	15	0.67	68.3%
France	69	25	0.46	12.2%

Major Projects Under Construction

Name	Maximum Capacity	Country	Construction started	Scheduled completion	Comments
Belo Monte Dam	11,181 MW	Brazil	March, 2011	2015	Preliminary construction underway. Construction suspended 14 days by court order Aug 2012
Siang Upper HE Project	11,000 MW	India	April, 2009	2024	Multi-phase construction over a period of 15 years. Construction was delayed due to dispute with China.
Tasang Dam	7,110 MW	Burma	March, 2007	2022	Controversial 228 meter tall dam with capacity to produce 35,446 GWh annually.
Xiangjiaba Dam	6,400 MW	China	November 26, 2006	2015	The last generator was commissioned on July 9, 2014

Grand Ethiopian Renaissance Dam	6,000 MW	Ethiopia	2011	2017	Located in the upper Nile Basin, drawing complaint from Egypt
Nuozhadu Dam	5,850 MW	China	2006	2017	
Jinping 2 Hydropower Station	4,800 MW	China	January 30, 2007	2014	To build this dam, 23 families and 129 local residents need to be moved. It works with Jinping 1 Hydropower Station as a group.
Diamer-Bhasha Dam	4,500 MW	Pakistan	October 18, 2011	2023	
Jinping 1 Hydropower Station	3,600 MW	China	November 11, 2005	2014	The sixth and final generator was commissioned on 15 July 2014
Jirau Power Station	3,300 MW	Brazil	2008	2013	Construction halted in March 2011 due to worker riots.
Guanyinyan Dam	3,000 MW	China	2008	2015	Construction of the roads and spillway started.
Lianghekou Dam	3,000 MW	China	2014	2023	
Dagangshan Dam	2,600 MW	China	August 15, 2008	2016	
Liyuan Dam	2,400 MW	China	2008	2013	
Tocoma Dam Bolívar State	2,160 MW	Venezuela	2004	2014	This power station would be the last development in the Low Caroni Basin, bringing the total to six power stations on the same river, including the 10,000MW Guri Dam.
Ludila Dam	2,100 MW	China	2007	2015	Brief construction halt in 2009 for environmental assessment.
Shuangjiangkou Dam	2,000 MW	China	December, 2007	2018	The dam will be 312 m high.
Ahai Dam	2,000 MW	China	July 27, 2006	2015	
Teles Pires Dam	1,820 MW	Brazil	2011	2015	
Site C Dam	1,100 MW	Canada	2015	2024	First large dam in western Canada since 1984
Lower Subansiri Dam	2,000 MW	India	2007	2016	

Run-of-the-river Hydroelectricity

Run-of-the-river hydroelectricity (ROR) is a type of hydroelectric generation plant whereby little or no water storage is provided. Run-of-the-river power plants may have no water storage at all or a limited amount of storage, in which case the storage reservoir is referred to as pondage. A plant without pondage has no water storage and is, therefore, subject to seasonal river flows. Thus, the plant will operate as an intermittent energy source while a plant with pondage can regulate the water flow at all times and can serve as a peaking power plant or base load power plant.

Chief Joseph Dam near Bridgeport, Washington, USA, is a major run-of-the-river station without a sizeable reservoir.

Concept

Run-of-the-river or ROR hydroelectricity is considered ideal for streams or rivers that can sustain a minimum flow or those regulated by a lake or reservoir upstream. A small dam is usually built to create a headpond ensuring that there is enough water entering the penstock pipes that lead to the turbines which are at a lower elevation. Projects with pondage, as opposed to those without pondage, can store water for daily load demands. In general, projects divert some or most of a river's flow (up to 95% of mean annual discharge) through a pipe and/or tunnel leading to electricity-generating turbines, then return the water back to the river downstream.

ROR projects are dramatically different in design and appearance from conventional hydroelectric projects. Traditional hydro dams store enormous quantities of water in reservoirs, sometimes flooding large tracts of land. In contrast, run-of-river projects do not have most of the disadvantages associated with dams and reservoirs, which is why they are often considered environmentally friendly.

Saint Marys Falls - run of the river (1902)

The use of the term "run-of-the-river" for power projects varies around the world. Some may consider a project ROR if power is produced with no water storage while limited storage is considered ROR by others. Developers may mislabel a project ROR to soothe public perception about its environmental or social effects. The Bureau of Indian Standards describes run-of-the-river hydroelectricity as:

A power station utilizing the run of the river flows for generation of power with sufficient pondage for supplying water for meeting diurnal or weekly fluctuations of demand. In such stations, the normal course of the river is not materially altered.

Many of the larger ROR projects have been designed to a scale and generating capacity rivaling some traditional hydro dams. For example, the Beauharnois Hydroelectric Generating Station in Quebec is rated at 1,853 MW and a 2006 proposal in British Columbia, Canada has been designed to generate 1027 megawatts capacity. Some run of the river projects are downstream of other dams and reservoirs. The run of the river project didn't build the reservoir, but does take advantage of the water supplied by it. An example would be the 1995, 1,436 MW La Grande-1 generating station, previous upstream dams and reservoirs are part of the 1980s James Bay Project.

Advantages

When developed with care to footprint size and location, ROR hydro projects can create sustainable energy minimizing impacts to the surrounding environment and nearby communities. Advantages include:

Cleaner Power, Fewer Greenhouse Gases

Like all hydro-electric power, run-of-the-river hydro harnesses the natural potential energy of water, eliminating the need to burn coal or natural gas to generate the electricity needed by consumers and industry.

Less Flooding/Reservoirs

Substantial flooding of the upper part of the river is not required for run-of-river projects as a large reservoir is not required. As a result, people living at or near the river don't need to be relocated and natural habitats and productive farmlands are not wiped out.

Disadvantages

"Unfirm" Power

Run-of-the-River power is considered an "unfirm" source of power: a run-of-the-river project has little or no capacity for energy storage and hence can't co-ordinate the output of electricity generation to match consumer demand. It thus generates much more power during times when seasonal river flows are high (i.e., spring freshet), and depending on location, much less during drier summer months or frozen winter months.

Availability of Sites

The potential power at a site is a result of the head and flow of water. By damming a river, the head is available to generate power at the face of the dam. Where a dam may create a reservoir hundreds of kilometers long, in run of the river the head is usually delivered by a canal, pipe or tunnel constructed upstream of the power house. Due to the cost of upstream construction, a steep drop in the river is desirable.

Environmental Impacts

Small, well-sited ROR projects can be developed with minimal environmental impacts. Larger projects have more environmental concerns. For example, Plutonic Power Corp.'s canceled Bute Inlet Hydroelectric Project in BC would have seen three clusters of run-of-river projects with 17 river diversions; as proposed, this run-of-river project would divert over 90 kilometres of streams and rivers into tunnels and pipelines, requiring 443 km of new transmission line, 267 km of permanent roads, and 142 bridges to be built in wilderness areas.

British Columbia's mountainous terrain and wealth of big rivers have made it a global testing ground for run-of-river technology. As of March 2010, there were 628 applications pending for new water licences solely for the purposes of power generation – representing more than 750 potential points of river diversion.

Concerns

- Diverting large amounts of river water reduces river flows, affecting water velocity and depth, minimizing habitat quality for fish and aquatic organisms; reduced flows can lead to excessively warm water for salmon and other fish in summer.

- New access roads and transmission lines can cause extensive habitat fragmentation for many species, allowing the introduction of invasive species and increases in undesirable human activities, like illegal hunting.

- The lack of reservoir storage may result in intermittent operation, reducing the project's viability.

Major Examples

- Belo Monte Dam, 11,233 megawatts (15,064,000 hp), Pará, Brazil

- Chief Joseph Dam, 2,620 megawatts (3,510,000 hp)

- Beauharnois Hydroelectric Power Station, 1,903 megawatts (2,552,000 hp)

- Bonneville Dam, 1,092 megawatts (1,464,000 hp)

- Satluj Jal Vldyut Nigam Ltd, Satluj River, Shimla, India, 1,500 megawatts (2,000,000 hp)

- Ghazi-Barotha Hydropower Project on River Indus in Pakistan, 1,450 megawatts (1,940,000 hp)

- La Grande-1 generating station, 1,436 megawatts (1,926,000 hp)

- Kohala Hydropower Project, Jhelum River, Muzaffarabad, Pakistan, 1,100 megawatts (1,500,000 hp)

- Neelum–Jhelum Hydropower Plant, Jhelum River, Muzaffarabad, Azad Kashmir, Pakistan, 969 megawatts (1,299,000 hp)

- Baglihar Hydroelectric Power Projecton Chenab River in India, 900 megawatts (1,200,000 hp)

- Carillon Generating Station, Quebec, Canada, 752 megawatts (1,008,000 hp)

- Upper Tamakoshi Project, Nepal, 456 MW

- Nyagak Hydroelectric Power Station on Nyagak River in Zombo District, Uganda, 3.5 megawatts (4,700 hp)

- East Toba/Montrose Hydro Project, British Columbia, Canada, 196 megawatts (263,000 hp)

- Forrest Kerr Hydro Project, British Columbia, Canada, 195 megawatts (261,000 hp)

- Patrind Hydropower Plant, Kunhar River, Pakistan, 150 megawatts (200,000 hp)

- Upper Toba Valley, British Columbia, Canada, 123 megawatts (165,000 hp)

- Upper Kotmale Hydropower Project (UKHP), Talawakele, Sri Lanka, 150 megawatts (200,000 hp)

- Sechelt Creek Generating Station, British Columbia, Canada, 16 megawatts (21,000 hp)

Small Hydro

Small 2MW hydro in Perthshire, Scotland

An 1895 hydroelectric plant near Telluride, Colorado.

Small hydro is the development of hydroelectric power on a scale serving a small community or industrial plant. The definition of a small hydro project varies, but a generating capacity of 1 to 20 megawatts (MW) is generally accepted, which aligns to the concept of distributed generation. The "small hydro" description may be stretched up to 50 MW in the United States, Canada and China. In contrast many hydroelectric projects are of enormous size, such as the generating plant at the Hoover Dam of 2,074 MW or the vast multiple projects of the Tennessee Valley Authority.

Small hydro can be further subdivided into mini hydro, usually defined as 100 to 1,000 kilowatts (kW), and micro hydro which is 5 to 100 kW. Micro hydro is usually the application of hydroelectric power sized for smaller communities, single families or small enterprise. The smallest installations are pico hydro, below 5 kW.

Small hydro plants may be connected to conventional electrical distribution networks as a source of low-cost renewable energy. Alternatively, small hydro projects may be built in isolated areas that would be uneconomic to serve from a network, or in areas where there is no national electrical distribution network. Since small hydro projects usually have minimal reservoirs and civil construction work, they are seen as having a relatively low environmental impact compared to large hydro. This decreased environmental impact depends strongly on the balance between stream flow and power production. One tool that helps evaluate this issue is the Flow Duration Curve or FDC. The FDC is a Pareto curve of a stream's daily flow rate vs. frequency. Reductions of diversion help the river's ecosystem, but reduce the hydro system's Return on Investment (ROI). The hydro system designer and site developer must strike a balance to maintain both the health of the stream and the economics.

Plants with reservoir, i.e. small storage and small pumped-storage hydropower plants, can contribute to distributed energy storage and decentralized peak and balancing electricity. Such plants can be built to integrate at the regional level intermittent renewable energy sources.

Growth

According to a report by REN21, during 2008 small hydro installations grew by 28% over year 2005 to raise the total world small hydro capacity to 85 gigawatts (GW). Over 70% of this was in China (with 65 GW), followed by Japan (3.5 GW), the United States (3 GW) and India (2 GW). China plans to electrify a further 10,000 villages by 2010 under their China Village Electrification Program using renewable energy, including further investments in small hydro and photovoltaics.

A 2013 report by the International Center on Small Hydro Power and UNIDO found that installed small hydro power around the globe was estimated at 75 GW and potential small hydro power was approximately 173 GW. China has made small hydro a priority and in 2010 had 45,000 installations, especially in rural areas, producing 160 Twh annually. Over 50% of the world's potential small hydro power was found in Asia however the report noted "It is possible in the future that more small hydropower potential might be identified both on the African and American continents".

In the mountains and rain forests of British Columbia, Canada there are a great many sites suitable for hydro development. However environmental concerns towards large reservoirs has halted large dam construction since the late 1980s. The solution was to offer guaranteed contracts to independent companies who have built 100 run of the river projects under 50MW. Power production without reservoirs varies dramatically though the year, but allows large dams to retain water and average out capacity. In 2014 they produced 18,000 GWh from 4,500 MW of capacity.

Generation

Hydroelectric power is the generation of electric power from the movement of water. A hydroelectric facility requires a dependable flow of water and a reasonable height of fall of water, called the head. In a typical installation, water is fed from a reservoir through a channel or pipe into a turbine. The pressure of the flowing water on the turbine blades causes the shaft to rotate. The rotating shaft is connected to an electrical generator which converts the motion of the shaft into electrical energy.

Historic small hydro with original equipment of 1920 in Ottenbach, Switzerland, still running for guided visits

Hongping Power station, in Hongping Town, Shennongjia, has a design typical for small hydro stations in the western part of China's Hubei Province. Water comes from the mountain behind the station, through the black pipe seen in the photo

Small hydro is often developed using existing dams or through development of new dams whose primary purpose is river and lake water-level control, or irrigation. Occasionally old, abandoned hydro sites may be purchased and re-developed, sometimes salvaging substantial parts of the installation such as penstocks and turbines, or sometimes just re-using the water rights associated with an abandoned site. Either of these cost saving advantages can make the ROI for a small hydro site well worth the use of existing site infrastructure & water rights.

Project Design

Many companies offer standardized turbine generator packages in the approximate size range of 200 kW to 10 MW. These "water to wire" packages simplify the planning and development of the site since one vendor looks after most of the equipment supply. Since non-recurring engineering costs are minimized and development cost is spread over multiple units, the cost of such systems is improved. While synchronous generators capable of isolated plant operation are often used, small hydro plants connected to an electrical grid system can use economical induction generators to further reduce installation cost and simplify control and operation.

Small hydro generating units commonly require a weir to form a headpond for diversion of inlet water to the turbine, unused water simply flows over the weir and there is no reservoir to store water for dry summers or frozen winters when generation may come to a halt. A preferred scenario is to have the inlet in an existing lake. Countries like India and China have policies in favor of small hydro, the regulatory process allows for building dams and reservoirs. In North America and Europe the regulatory process is too long and expensive to consider having a dam and a reservoir for a small project.

Other small hydro schemes may use tidal energy or propeller-type turbines immersed in flowing water to extract energy. Tidal schemes may require water storage or electrical energy storage to level out the intermittent (although exactly predictable) flow of power.

Small hydro projects usually have faster environmental and licensing procedures, and since the equipment is usually in serial production, standardized and simplified, and since the civil works construction is also reduced, small hydro projects may be developed very rapidly. The physically smaller size of equipment makes it easier to transport to remote areas without good road or rail access.

Sample List of Small Installations Worldwide

Africa

- Zengamina, a 700 kW plant in Kalene Hill, Mwinilunga District in northwestern Zambia. 2008

Asia

- Meenvallam Small Hydroelectric Project, Palakkad district, Kerala, India.

- Bario Asal & Arur Layun Micro-Hydro Community Project, Kelabit Highlands, Sarawak, Malaysia. 2009

- Cantigas Mini Hydro Electric Power Plant, San Fernando, Romblon, Philippines.

- Pakpattan Hydro Power Project, a 2.82MW plant on Pakpattan Cannal in Pakpattan District, Pakistan 2015

Europe

- Green Valleys Project, Brecon Beacons National Park, Wales, United Kingdom. Joint winner of £300K in Big Green Challenge 2009.

- St Catherine's, a National Trust site near Windermere, Westmorland, United Kingdom.

- North America

- Ames Hydroelectric Generating Plant, Colorado, United States. On the List of IEEE Milestones

- Childs-Irving Hydroelectric Facilities, Arizona, United States. 1916 Now decommissioned

- Snoqualmie Falls, Washington, United States. 1957

- Malibu Hydro, British Columbia, Canada. 2005

- Cloudworks Energy has built several projects in western Canada

Micro Hydro

Micro hydro in northwest Vietnam

Micro hydro is a type of hydroelectric power that typically produces from 5 kW to 100 kW of electricity using the natural flow of water. Installations below 5 kW are called pico hydro. These installations can provide power to an isolated home or small community, or are sometimes connected to electric power networks, particularly where net metering is offered. There are many of these installations around the world, particularly in developing nations as they can provide an economical source of energy without the purchase of fuel. Micro hydro systems complement solar PV power systems because in many areas, water flow, and thus available hydro power, is highest in the winter when solar energy is at a minimum. Micro hydro is frequently accomplished with a pelton wheel for high head, low flow water supply. The installation is often just a small dammed pool, at the top of a waterfall, with several hundred feet of pipe leading to a small generator housing. maot kag nawong

Construction

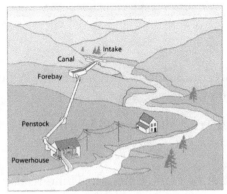

In this microhydropower system, water is diverted into the penstock. Some generators can be placed directly into the stream.

Typical microhydro setup.

Construction details of a microhydro plant are site-specific. Sometimes an existing mill-pond or other artificial reservoir is available and can be adapted for power production. In general, microhydro systems are made up of a number of components. The most important include the intake where water is diverted from the natural stream, river, or perhaps a waterfall. An intake structure such as a catch box is required to screen out floating debris and fish, using a screen or array of bars to keep out large objects. In temperate climates this structure must resist ice as well. The intake may have a gate to allow the system to be dewatered for inspection and maintenance.

The intake then tunnels water through a pipeline (penstock) to the powerhouse building containing a turbine. In mountainous areas, access to the route of the penstock may provide considerable challenges. If the water source and turbine are far apart, the construction of the penstock may be the largest part of the costs of construction. At the turbine, a controlling valve is installed to regulate the flow and the speed of the turbine. The turbine converts the flow and pressure of the water to mechanical energy; the water emerging from the turbine returns to the natural watercourse along a tailrace channel. The turbine turns a generator, which is then connected to electrical loads; this might be directly connected to the power system of a single building in very small installations, or may be connected to a community distribution system for several homes or buildings.

Usually microhydro installations do not have a dam and reservoir, like large hydroelectric plants have, relying on a minimal flow of water to be available year-round.

Head and Flow Characteristics

Microhydro systems are typically set up in areas capable of producing up to 100 kilowatts of electricity. This can be enough to power a home or small business facility. This production range is calculated in terms of "head" and "flow". The higher each of these are, the more power available. Hydraulic head is the pressure measurement of water falling in a pipe expressed as a function of the vertical distance the water falls. This change in elevation is usually measured in feet or meters. A drop of at least 2 feet is required or the system may not be feasible. When quantifying head, both gross and net head must be considered. Gross head approximates power accessibility through the vertical distance measurement alone whereas net head subtracts pressure lost due to friction in piping from the gross head. "Flow" is the actual quantity of water falling from a site and is usually measured in gallons per minute, cubic feet per second, or liters per second. Low flow/high head installations in steep terrain have significant pipe costs. A long penstock starts with low pressure pipe at the top and progressively higher pressure pipe closer to the turbine in order to reduce pipe costs.

Power from such a system can be calculated by the equation $P=Q*H/k$, where Q is the flow rate in gallons per minute, H is the static head, and k is a constant of 5,310 gal*ft/min*kW. For instance, for a system with a flow of 500 gallons per minute and a static head of 60 feet, the theoretical maximum power output is 5.65 kW. The system is prevented from 100% efficiency (from obtaining all 5.65 kW) due to the real world, such as: turbine efficiency, friction in pipe, and conversion from potential to kinetic energy. Turbine efficiency is generally between 50-80%, and pipe friction is accounted for using the Hazen–Williams equation.

Regulation and Operation

Typically, an automatic controller operates the turbine inlet valve to maintain constant speed (and frequency) when the load changes on the generator. In a system connected to a grid with multiple sources, the turbine control ensures that power always flows out from the generator to the system.

The frequency of the alternating current generated needs to match the local standard utility frequency. In some systems, if the useful load on the generator is not high enough, a load bank may be automatically connected to the generator to dissipate energy not required by the load; while this wastes energy, it may be required if it's not possible to control the water flow through the turbine.

An induction generator always operates at the grid frequency irrespective of its rotation speed; all that is necessary is to ensure that it is driven by the turbine faster than the synchronous speed so that it generates power rather than consuming it. Other types of generator can use a speed control systems for frequency matching.

With the availability of modern power electronics it is often easier to operate the generator at an arbitrary frequency and feed its output through an inverter which produces output at grid frequency. Power electronics now allow the use of permanent magnet alternators that produce wild AC to be stabilised. This approach allows low speed / low head water turbines to be competitive; they can run at the best speed for extraction of energy, and the power frequency is controlled by the electronics instead of the generator.

Very small installations (pico hydro), a few kilowatts or smaller, may generate direct current and charge batteries for peak use times.

Turbine Types

Several types of water turbines can be used in micro hydro installations, selection depending on the head of water, the volume of flow, and such factors as availability of local maintenance and transport of equipment to the site. For mountainous regions where a waterfall of 50 meters or more may be available, a Pelton wheel can be used. For low head installations, Francis or propeller-type turbines are used. Very low head installations of only a few meters may use propeller-type turbines in a pit. The very smallest micro hydro installations may successfully use industrial centrifugal pumps, run in reverse as prime movers; while the efficiency may not be as high as a purpose-built runner, the relatively low cost makes the projects economically feasible.

In low-head installations, maintenance and mechanism costs often become important. A low-head system moves larger amounts of water, and is more likely to encounter surface debris. For this reason a Banki turbine also called Ossberger turbine, a pressurized self-cleaning crossflow waterwheel, is often preferred for low-head microhydropower systems. Though less efficient, its simpler structure is less expensive than other low-head turbines of the same capacity. Since the water flows in, then out of it, it cleans itself and is less prone to jam with debris.

- Screw turbine (Reverse Archimedes' screw): two low-head schemes in England, Settle Hydro and Torrs Hydro use an Archimedes' screw which is another debris-tolerant design. Efficiency 85%.

- Gorlov: the Gorlov helical turbine free stream or constrained flow with or without a dam,

- Francis and propeller turbines.

- Kaplan turbine : an alternative to the traditional kaplan turbine is a large diameter, slow turning, permanent magnet, sloped open flow VLH turbine with efficiencies of 90%.

- Water wheel : advanced hydraulic water wheels and hydraulic wheel-part reaction turbine can have hydraulic efficiencies of 67% and 85% respectively.

- Gravitation water vortex power plant : part of the river flow at a weir or natural water fall is diverted into a round basin with a central bottom exit that creates a vortex. A simple rotor (and connected generator) is moved by the kinetic energy. Efficiencies of 83% down to 64% at 1/3 part flow.

Use

Microhydro systems are very flexible and can be deployed in a number of different environments. They are dependent on how much water flow the source (creek, river, stream) has and the velocity of the flow of water. Energy can be stored in battery banks at sites that are far from a facility or used in addition to a system that is directly connected so that in times of high demand there is additional reserve energy available. These systems can be designed to minimize community and environmental impact regularly caused by large dams or other mass hydroelectric generation sites.

Potential for Rural Development

In relation to rural development, the simplicity and low relative cost of micro hydro systems open up new opportunities for some isolated communities in need of electricity. With only a small stream needed, remote areas can access lighting and communications for homes, medical clinics, schools, and other facilities. Microhydro can even run a certain level of machinery supporting small businesses. Regions along the Andes mountains and in Sri Lanka and China already have similar, active programs. One seemingly unexpected use of such systems in some areas is to keep young community members from moving into more urban regions in order to spur economic growth. Also, as the possibility of financial incentives for less carbon intensive processes grows, the future of microhydro systems may become more appealing.

Micro-hydro installations can also provide multiple uses. For instance, micro-hydro projects in rural Asia have incorporated agro-processing facilities such as rice mills – alongside standard electrification – into the project design.

Cost

The cost of a micro hydro plant can be between 1,000 and 20,000 U.S. dollars

Advantages and Disadvantages

System Advantages

Microhydro power is generated through a process that utilizes the natural flow of water. This power is most commonly converted into electricity. With no direct emissions resulting from this conversion process, there are little to no harmful effects on the environment, if planned well, thus supplying power from a renewable source and in a sustainable manner. Microhydro is considered a "run-of-river" system meaning that water diverted from the stream or river is redirected back into the same watercourse. Adding to the potential economic benefits of microhydro is efficiency, reliability, and cost effectiveness.

System Disadvantages

Microhydro systems are limited mainly by characteristics of the site. The most direct limitation comes from small sources with minuscule flow. Likewise, flow can fluctuate seasonally in some areas. Lastly, though perhaps the foremost disadvantage is the distance from the power source to the site in need of energy. This distributional issue as well as the others are key when considering using a microhydro system.

Conduit Hydroelectricity

Conduit hydroelectricity (or conduit hydropower) is a method of using mechanical energy of water as part of the water delivery system through man-made conduits to generate electricity. Generally, the conduits are existing water pipelines such as in public water supply. Some definitions expand the definition of conduits to be existing tunnels, canals, or aqueducts that are used primarily for other water delivery purposes than electricity generation.

Historically, electricity generation from water pipelines was rare because the water would have been pumped by other engines in the system prior to the intake of water turbines to generate electricity. The energy generated from the turbines would have been offset by the power used in pumping, canceling out the power generation benefit. However, there have been renewed interests to apply this method to recover energy when there is a need to reduce pressure in the water supply system that is normally done through pressure reducing valves. The conduit hydroelectricity generation in this case can be done by replacing the pressure reducing valves with small turbines and electrical generators.

Pico Hydro

A pico hydro system made by the *Sustainable Vision project* from Baylor University

Pico hydro is a term used for hydroelectric power generation of under 5 kW. It is useful in small, remote communities that require only a small amount of electricity – for example, to power one or two fluorescent light bulbs and a TV or radio in 50 or so homes. Even smaller turbines of 200–300 W may power a single home in a developing country with a drop of only one meter. Pico-hydro setups typically are run-of-stream, meaning that a reservoir of water is not created, only a small weir is common, pipes divert some of the flow, drop this down a gradient, and through the turbine before being exhausted back to the stream.

Like other hydroelectric and renewable source power generation, pollution and consumption of fossil fuels is reduced, though there is still typically an environmental cost to the manufacture of the generator and distribution methods.

Small-scale DIY Hydroplants

With a growing DIY-community and an increasing interest in environmentally friendly "green energy", some hobbyists have endeavored to build their own hydroelectric plants from old water mills, or from kits or from scratch. The DIY-community has used abandoned water mills to mount a waterwheel and electrical generating components. This approach has also been popularised in the TV-series *It's Not Easy Being Green*. These are usually smaller turbines of ~5 kW or less. Through the internet, the community is now able to obtain plans to construct DIY-water turbines, and there is a growing trend toward building them for domestic requirements. The DIY-hydroelectric plants are now being used both in developed countries and in developing countries, to power residences and small businesses. Two examples of pico hydro power can be found in the towns of Kithamba and Thimba in the Central Province of Kenya. These produce 1.1 kW and 2.2 kW, respectively. Local residents were trained to maintain the hydro schemes. The pico hydro sites in Kenya won Ashden Awards for Sustainable Energy.

Manufacturers

In Vietnam, several Chinese manufacturers have sold pico-powerplants at prices as low as $20–70 for a powerplant of 300–500 W. However, the devices sold are said to be low in quality and may damage connected equipment if connected improperly.

Sam Redfield of the Appropriate Infrastructure Development Group (AIDG) has developed a pico-hydro generator made from common PVC pipe and a modified Toyota alternator housed in a five gallon bucket. The generator was developed to provide power to communities without access to the electricity grid in developing countries. Envisioned as an energy source to charge cell phones, provide lighting and charge batteries, the generator is designed to be made by artisans with basic skills and can be built for less than $150.00. The Toyota alternator used in the generator is converted to a permanent magnet alternator allowing it to generate power at low RPMs. The Five Gallon Bucket Hydroelectric Generator was the subject of a work group at the 2008 International Development Design Summit (IDDS) at the Massachusetts Institute of Technology. During the Summer of 2013 an energy project in Abra Malaga, Peru was completed using the bucket generator.

A website has been put together as a forum for ideas and further iterations of the generator that includes a *build manual*. Common centrifugal water pumps, can be operated in reverse to act as turbines. While these machines rarely have optimum hydraulic characteristics when operated as turbines, their availability and low cost makes them attractive.

Transmission Distance

If the power will be used more than 100 feet from the generator, then the transmission distance may be an important consideration. Many small systems use automotive alternators producing 12 VDC, and possibly charging a battery. For example, a 12 V system that produces 1 kW of power has a flow of 80 A and the wire size is 4-gauge. The cost to run two strands of wire 1000' is $2400(US). To avoid such a large wire cost a higher voltage and lower amperage is required. If a 240 VAC alternator is used instead the flow is only 4 A over 1000' of 18 gauge wire costing $180(US). The cost of wire resulted in North America using 120/240 VAC after DC voltage lost the War of Currents in the late 1800s. Another approach to reduce wire costs is to have a 12 VDC alternator with a short high amperage connection to an inverter outputting 120 VAC or 240 VAC at a much lower amperage on a long length of thinner wire.

Power Station

The Athlone Power Station in Cape Town, South Africa

Hydroelectric power station at Gabčíkovo Dam, Slovakia

A power station, also referred to as a generating station, power plant, powerhouse, or generating plant, is an industrial facility for the generation of electric power. Most power stations contain one or more generators, a rotating machine that converts mechanical power into electrical power. The relative motion between a magnetic field and a conductor creates an electrical current. The energy source harnessed to turn the generator varies widely. Most power stations in the world burn fossil fuels such as coal, oil, and natural gas to generate electricity. Others use nuclear power, but there is an increasing use of cleaner renewable sources such as solar, wind, wave and hydroelectric.

History

In 1868 a hydro electric power station was designed and built by Lord Armstrong at Cragside, England. It used water from lakes on his estate to power Siemens dynamos. The electricity supplied power to lights, heating, produced hot water, ran an elevator as well as labor-saving devices and farm buildings.

In the early 1870s Belgian inventor Zénobe Gramme invented a generator powerful enough to produce power on a commercial scale for industry.

In the autumn of 1882, a central station providing public power was built in Godalming, England. It was proposed after the town failed to reach an agreement on the rate charged by the gas company, so the town council decided to use electricity. It used hydroelectric power that was used to street and household lighting. The system was not a commercial success and the town reverted to gas.

In 1882 a public power station, the Edison Electric Light Station, was built in London, was a project of Thomas Edison organized by Edward Johnson. A Babcock & Wilcox boiler powered a 125-horse-power steam engine that drove a 27-ton generator. This supplied electricity to premises in the area that could be reached through the culverts of the viaduct without digging up the road, which was the monopoly of the gas companies. The customers included the City Temple and the Old Bailey. Another important customer was the Telegraph Office of the General Post Office, but this could not be reached though the culverts. Johnson arranged for the supply cable to be run overhead, via Holborn Tavern and Newgate.

In September 1882 in New York, the Pearl Street Station was established by Edison to provide electric lighting in the lower Manhattan Island area. The station ran until destroyed by fire in 1890. The station used reciprocating steam engines to turn direct-current generators. Because of the DC distribution, the service area was small, limited by voltage drop in the feeders. The War of Currents eventually resolved in favor of AC distribution and utilization, although some DC systems persisted to the end of the 20th century. DC systems with a service radius of a mile (kilometer) or so were necessarily smaller, less efficient of fuel consumption, and more labor-intensive to operate than much larger central AC generating stations.

AC systems used a wide range of frequencies depending on the type of load; lighting load using higher frequencies, and traction systems and heavy motor load systems preferring lower frequencies. The economics of central station generation improved greatly when unified light and power systems, operating at a common frequency, were developed. The same generating plant that fed large industrial loads during the day, could feed commuter railway systems during rush hour and then serve lighting load in the evening, thus improving the system load factor and reducing the cost of electrical energy overall. Many exceptions existed, generating stations were dedicated to power or light by the choice of frequency, and rotating frequency changers and rotating converters were particularly common to feed electric railway systems from the general lighting and power network.

Throughout the first few decades of the 20th century central stations became larger, using higher steam pressures to provide greater efficiency, and relying on interconnections of multiple generating stations to improve reliability and cost. High-voltage AC transmission allowed hydroelectric power to be conveniently moved from distant waterfalls to city markets. The advent of the steam turbine in central station service, around 1906, allowed great expansion of generating capacity. Generators were no longer limited by the power transmission of belts or the relatively slow speed of reciprocating engines, and could grow to enormous sizes. For example, Sebastian Ziani de Ferranti planned what would have been the largest reciprocating steam engine ever built for a proposed new central station, but scrapped the plans when turbines became available in the necessary size. Building power systems out of central stations required combinations of engineering skill and financial acumen in equal measure. Pioneers of central station generation include George Westinghouse and Samuel Insull in the United States, Ferranti and Charles Hesterman Merz in UK, and many others.

Thermal Power Stations

Rotor of a modern steam turbine, used in a power station

In thermal power stations, mechanical power is produced by a heat engine that transforms thermal energy, often from combustion of a fuel, into rotational energy. Most thermal power stations produce steam, so they are sometimes called steam power stations. Not all thermal energy can be transformed into mechanical power, according to the second law of thermodynamics; therefore, there is always heat lost to the environment. If this loss is employed as useful heat, for industrial processes or district heating, the power plant is referred to as a cogeneration power plant or CHP (combined heat-and-power) plant. In countries where district heating is common, there are dedicated heat plants called heat-only boiler stations. An important class of power stations in the Middle East uses by-product heat for the desalination of water.

The efficiency of a thermal power cycle is limited by the maximum working fluid temperature produced. The efficiency is not directly a function of the fuel used. For the same steam conditions, coal-, nuclear- and gas power plants all have the same theoretical efficiency. Overall, if a system is on constantly (base load) it will be more efficient than one that is used intermittently (peak load). Steam turbines generally operate at higher efficiency when operated at full capacity.

Besides use of reject heat for process or district heating, one way to improve overall efficiency of a power plant is to combine two different thermodynamic cycles in a combined cycle plant. Most commonly, exhaust gases from a gas turbine are used to generate steam for a boiler and a steam turbine. The combination of a "top" cycle and a "bottom" cycle produces higher overall efficiency than either cycle can attain alone.

Classification

St. Clair Power Plant, a large coal-fired generating station in Michigan, United States

Ikata Nuclear Power Plant, Japan

By Heat Source

- Fossil-fuel power stations may also use a steam turbine generator or in the case of natural gas-fired plants may use a combustion turbine. A coal-fired power station produces heat by burning coal in a steam boiler. The steam drives a steam turbine and generator that then produces electricity. The waste products of combustion include ash, sulphur dioxide, nitrogen oxides and carbon dioxide. Some of the gases can be removed from the waste stream to reduce pollution.

- Nuclear power plants use a nuclear reactor's heat that is transferred to steam which then operates a steam turbine and generator. About 20 percent of electric generation in the USA is produced by nuclear power plants.

- Geothermal power plants use steam extracted from hot underground rocks. These rocks are heated by the decay of radioactive material in the Earth's crust.

- Biomass-fuelled power plants may be fuelled by waste from sugar cane, municipal solid waste, landfill methane, or other forms of biomass.

- In integrated steel mills, blast furnace exhaust gas is a low-cost, although low-energy-density, fuel.

- Waste heat from industrial processes is occasionally concentrated enough to use for power generation, usually in a steam boiler and turbine.

- Solar thermal electric plants use sunlight to boil water and produce steam which turns the generator.

By Prime Mover

- Steam turbine plants use the dynamic pressure generated by expanding steam to turn the blades of a turbine. Almost all large non-hydro plants use this system. About 90 percent of all electric power produced in the world is through use of steam turbines.

- Gas turbine plants use the dynamic pressure from flowing gases (air and combustion products) to directly operate the turbine. Natural-gas fuelled (and oil fueled) combustion turbine plants can start rapidly and so are used to supply "peak" energy during periods of high demand, though at higher cost than base-loaded plants. These may be comparatively small units, and sometimes completely unmanned, being remotely operated. This type was pioneered by the UK, Princetown being the world's first, commissioned in 1959.

- Combined cycle plants have both a gas turbine fired by natural gas, and a steam boiler and steam turbine which use the hot exhaust gas from the gas turbine to produce electricity. This greatly increases the overall efficiency of the plant, and many new baseload power plants are combined cycle plants fired by natural gas.

- Internal combustion reciprocating engines are used to provide power for isolated communities and are frequently used for small cogeneration plants. Hospitals, office buildings, industrial plants, and other critical facilities also use them to provide backup power in case of a power outage. These are usually fuelled by diesel oil, heavy oil, natural gas, and landfill gas.

- Microturbines, Stirling engine and internal combustion reciprocating engines are low-cost solutions for using opportunity fuels, such as landfill gas, digester gas from water treatment plants and waste gas from oil production.

By Duty

Power plants that can be dispatched (scheduled) to provide energy to a system include:

- Base load power plants run nearly continually to provide that component of system load that doesn't vary during a day or week. Baseload plants can be highly optimized for low fuel cost, but may not start or stop quickly during changes in system load. Examples of base-load plants would include large modern coal-fired and nuclear generating stations, or hydro plants with a predictable supply of water.

- Peaking power plants meet the daily peak load, which may only be for one or two hours each day. While their incremental operating cost is always higher than base load plants, they are required to ensure security of the system during load peaks. Peaking plants include simple cycle gas turbines and sometimes reciprocating internal combustion engines, which can be started up rapidly when system peaks are predicted. Hydroelectric plants may also be designed for peaking use.

- Load following power plants can economically follow the variations in the daily and weekly load, at lower cost than peaking plants and with more flexibility than baseload plants.

Non-dispatchable plants include such sources as wind and solar energy; while their long-term contribution to system energy supply is predictable, on a short-term (daily or hourly) base their energy must be used as available since generation cannot be deferred. Contractual arrangements ("take or pay") with independent power producers or system interconnections to other networks may be effectively non-dispatchable.

Cooling Towers

Cooling towers showing evaporating water at Ratcliffe-on-Soar Power Station, United Kingdom

"Camouflaged" natural draft wet cooling tower

All thermal power plants produce waste heat energy as a byproduct of the useful electrical energy produced. The amount of waste heat energy equals or exceeds the amount of energy converted into useful electricity. Gas-fired power plants can achieve as much as 65 percent conversion efficiency, while coal and oil plants achieve around 30 to 49 percent. The waste heat produces a temperature rise in the atmosphere, which is small compared to that produced by greenhouse-gas emissions from the same power plant. Natural draft wet cooling towers at many nuclear power plants and large fossil fuel-fired power plants use large hyperboloid chimney-like structures (as seen in the image at the right) that release the waste heat to the ambient atmosphere by the evaporation of water.

However, the mechanical induced-draft or forced-draft wet cooling towers in many large thermal power plants, nuclear power plants, fossil-fired power plants, petroleum refineries, petrochemical plants, geothermal, biomass and waste-to-energy plants use fans to provide air movement upward through downcoming water, and are not hyperboloid chimney-like structures. The induced or forced-draft cooling towers are typically rectangular, box-like structures filled with a material that enhances the mixing of the upflowing air and the downflowing water.

In areas with restricted water use, a dry cooling tower or directly air-cooled radiators may be necessary, since the cost or environmental consequences of obtaining make-up water for evaporative cooling would be prohibitive. These coolers have lower efficiency and higher energy consumption to drive fans, compared to a typical wet, evaporative cooling tower.

Once-through Cooling Systems

Electric companies often prefer to use cooling water from the ocean, a lake, or a river, or a cooling pond, instead of a cooling tower. This single pass or once-through cooling system can save the cost of a cooling tower and may have lower energy costs for pumping cooling water through the plant's heat exchangers. However, the waste heat can cause thermal pollution as the water is discharged. Power plants using natural bodies of water for cooling are designed with mechanisms such as fish screens, to limit intake of organisms into the cooling machinery. These screens are only partially effective and as a result billions of fish and other aquatic organisms are killed by power plants each year. For example, the cooling system at the Indian Point Energy Center in New York kills over a billion fish eggs and larvae annually.

A further environmental impact is that aquatic organisms which adapt to the warmer discharge water may be injured if the plant shuts down in cold weather.

Water consumption by power stations is a developing issue.

In recent years, recycled wastewater, or grey water, has been used in cooling towers. The Calpine Riverside and the Calpine Fox power stations in Wisconsin as well as the Calpine Mankato power station in Minnesota are among these facilities.

Power from Renewable Energy

Power stations can also generate electrical energy from renewable energy sources.

Hydroelectric Power Station

Three Gorges Dam, Hubei, China

In a hydroelectric power station water flows though turbines using hydropower to generate hydroelectricity. Power is captured from the gravitational force of water falling through penstocks to water turbines connected to generators. The amount of power available is a combination of height and flow. A wide range of Dams may be built to raise the water level, and create a lake for storing water. Hydropower is produced in 150 countries, with the Asia-Pacific region generating 32 percent of global hydropower in 2010. China is the largest hydroelectricity producer, with 721 terawatt-hours of production in 2010, representing around 17 percent of domestic electricity use.

Pumped Storage

A pumped-storage is a reversible hydroelectric power plant. They are a net consumer of energy but can be used for storage to smooth peaks and troughs in overall electricity demand. Pumped storage plants typically use "spare" electricity during off peak periods to pump freshwater or saltwater from a lower reservoir to an upper reservoir. Because the pumping takes place "off peak", electricity is typically cheaper than at peak times. This is because power sources such as coal-fired, solar and wind are not switched off and remain in service even when demand is low. During hours of peak demand, when the electricity price is high, the water pumped to the upper reservoir is allowed to flow back to the lower reservoir through a water turbine connected to an electricity generator. Unlike coal power stations, which can take more than 12 hours to start up from cold, the hydroelectric plant can be brought into service in a few minutes, ideal to meet a peak load demand. Two substantial pumped storage schemes are in South Africa, Palmiet Pumped Storage Scheme and another in the Drakensberg, Ingula Pumped Storage Scheme.

Solar

Nellis Solar Power Plant in Nevada, United States

Solar energy can be turned into electricity either directly in solar cells, or in a concentrating solar power plant by focusing the light to run a heat engine.

A solar photovoltaic power plant converts sunlight into direct current electricity using the photoelectric effect. Inverters change the direct current into alternating current for connection to the electrical grid. This type of plant does not use rotating machines for energy conversion.

Solar thermal power plants are another type of solar power plant. They use either parabolic troughs or heliostats to direct sunlight onto a pipe containing a heat transfer fluid, such as oil. The heated oil is then used to boil water into steam, which turns a turbine that drives an electrical generator. The central tower type of solar thermal power plant uses hundreds or thousands of mirrors, depending on size, to direct sunlight onto a receiver on top of a tower. Again, the heat is used to produce steam to turn turbines that drive electrical generators.

Wind

Wind turbines can be used to generate electricity in areas with strong, steady winds, sometimes offshore. Many different designs have been used in the past, but almost all modern turbines being produced today use a three-bladed, upwind design. Grid-connected wind turbines now being built are much larger than the units installed during the 1970s. They thus produce power more cheaply and reliably than earlier models. With larger turbines (on the order of one megawatt), the blades move more slowly than older, smaller, units, which makes them less visually distracting and safer for birds.

Wind turbines in Texas, United States

Marine

Marine energy or marine power (also sometimes referred to as ocean energy or ocean power) refers to the energy carried by ocean waves, tides, salinity, and ocean temperature differences. The movement of water in the world's oceans creates a vast store of kinetic energy, or energy in motion. This energy can be harnessed to generate electricity to power homes, transport and industries.

The term marine energy encompasses both wave power — power from surface waves, and tidal power — obtained from the kinetic energy of large bodies of moving water. Offshore wind power is not a form of marine energy, as wind power is derived from the wind, even if the wind turbines are placed over water.

The oceans have a tremendous amount of energy and are close to many if not most concentrated populations. Ocean energy has the potential of providing a substantial amount of new renewable energy around the world.

Osmosis

Osmotic Power Prototype at Tofte (Hurum), Norway

Salinity gradient energy is called pressure-retarded osmosis. In this method, seawater is pumped into a pressure chamber that is at a pressure lower than the difference between the pressures of saline water and fresh water. Freshwater is also pumped into the pressure chamber through a membrane, which increases both the volume and pressure of the chamber. As the pressure differences are compensated, a turbine is spun creating energy. This method is being specifically studied by the Norwegian utility Statkraft, which has calculated that up to 25 TWh/yr would be available from this process in Norway. Statkraft has built the world's first prototype osmotic power plant on the Oslo fiord which was opened on November 24, 2009.

Biomass

Metz biomass power station

Biomass energy can be produced from combustion of waste green material to heat water into steam and drive a steam turbine. Bioenergy can also be processed through a range of temperatures and pressures in gasification, pyrolysis or torrefaction reactions. Depending on the desired end product, these reactions create more energy-dense products (syngas, wood pellets, biocoal) that can then be fed into an accompanying engine to produce electricity at a much lower emission rate when compared with open burning.

Storage Power Stations

It is possible to store energy and produce the electricity at a later time like in Pumped-storage hydroelectricity, Thermal energy storage, Flywheel energy storage, Battery storage power station and so on.

Typical Power Output

The power generated by a power station is measured in multiples of the watt, typically megawatts (10^6 watts) or gigawatts (10^9 watts). Power stations vary greatly in capacity depending on the type of power plant and on historical, geographical and economic factors. The following examples offer a sense of the scale.

Many of the largest operational onshore wind farms are located in the USA. As of 2011, the Roscoe Wind Farm is the second largest onshore wind farm in the world, producing 781.5 MW of power, followed by the Horse Hollow Wind Energy Center (735.5 MW). As of July 2013, the London Array in United Kingdom is the largest offshore wind farm in the world at 630 MW, followed by Thanet Offshore Wind Project in United Kingdom at 300 MW.

As of 2015, the largest photovoltaic (PV) power plants in the world are led by Longyangxia Dam Solar Park in China, rated at 850 megawatts.

Solar thermal power stations in the U.S. have the following output:

The country's largest solar facility at Kramer Junction has an output of 354 MW

The Blythe Solar Power Project planned production is estimated at 485 MW

Aerial view of the Three Mile Island Nuclear Generating Station, USA

Large coal-fired, nuclear, and hydroelectric power stations can generate hundreds of megawatts to multiple gigawatts. Some examples:

The Three Mile Island Nuclear Generating Station in the USA has a rated capacity of 802 megawatts.

The coal-fired Ratcliffe-on-Soar Power Station in the UK has a rated capacity of 2 gigawatts.

The Aswan Dam hydro-electric plant in Egypt has a capacity of 2.1 gigawatts.

The Three Gorges Dam hydro-electric plant in China has a capacity of 22.5 gigawatts.

Gas turbine power plants can generate tens to hundreds of megawatts. Some examples:

The Indian Queens simple-cycle peaking power station in Cornwall UK, with a single gas turbine is rated 140 megawatts.

The Medway Power Station, a combined-cycle power station in Kent, UK with two gas turbines and one steam turbine, is rated 700 megawatts.

The rated capacity of a power station is nearly the maximum electrical power that that power station can produce. Some power plants are run at almost exactly their rated capacity all the time, as a non-load-following base load power plant, except at times of scheduled or unscheduled maintenance.

However, many power plants usually produce much less power than their rated capacity.

In some cases a power plant produces much less power than its rated capacity because it uses an intermittent energy source. Operators try to pull maximum available power from such power plants, because their marginal cost is practically zero, but the available power varies widely—in particular, it may be zero during heavy storms at night.

In some cases operators deliberately produce less power for economic reasons. The cost of fuel to run a load following power plant may be relatively high, and the cost of fuel to run a peaking power plant is even higher—they have relatively high marginal costs. Operators keep power plants turned off ("operational reserve") or running at minimum fuel consumption ("spinning reserve") most of the time. Operators feed more fuel into load following power plants only when the demand rises above what lower-cost plants (i.e., intermittent and base load plants) can produce, and then feed more fuel into peaking power plants only when the demand rises faster than the load following power plants can follow.

Operations

Control room of a power plant

The term Power station is generally limited to those able to be despatched by a system operator (i.e. the system operator can, by one means or another, alter the planned output of the generating facility).

The power station operator has several duties in the electricity-generating facility. Operators are responsible for the safety of the work crews that frequently do repairs on the mechanical and electrical equipment. They maintain the equipment with periodic inspections and log temperatures, pressures and other important information at regular intervals. Operators are responsible for starting and stopping the generators depending on need. They are able to synchronize and adjust the voltage output of the added generation with the running electrical system, without upsetting the system. They must know the electrical and mechanical systems in order to troubleshoot solve/fix problems in the facility and add to the reliability of the facility. Operators must be able to respond to an emergency and know the procedures in place to deal with it.

Underground Power Station

An underground power station is a type of hydroelectric power station constructed by excavating the major components (e.g. machine hall, penstocks, and tailrace) from rock, rather than the more common surface-based construction methods.

One or more conditions impact whether a power station is constructed underground. The terrain or geology around a dam is taken into consideration as gorges or steep valleys may not accommodate a surface power station. A power station within bedrock may be more inexpensive to construct than a surface power station on loose soil. Avalanche-prone valleys often make a surface station unfeasible as well. After World War II, large hydroelectric power stations were placed underground more often in order to protect them from airstrikes.

Inside the Robert-Bourassa generating station, in northern Quebec, the world's largest underground power station, with an installed capacity of 5,616 MW.

Often underground power stations form part of pumped storage hydroelectricity schemes, whose basic function is to level load: they use cheap or surplus off-peak power to pump water from a lower lake to an upper lake, then, during peak periods (when electricity prices are often high), the power station generates power from the water held in the upper lake.

Notable Examples

Some notable underground power stations are:

- Snoqualmie Falls Hydroelectric Plant in King County, Washington, United States, built in two stages, Plant 1, completed in 1899 was the world's first completely underground power station and is still used to provide power to the Seattle area. The two power houses have a combined installed capacity of 53.9 MW.

- Chaira Hydro Power Plant, Bulgaria, is the largest underground power station in the Balkans, built from 1980 to 1998. It has an installed capacity of 864 MW from four 216 MW reversible Francis turbines with a net rated head of 2,300 feet (701 m), and maximal speed of 600 rpm.

- Churchill Falls Generating Station, Newfoundland and Labrador, Canada is the second largest underground power station in the world. It generates 5,428 MW from 11 turbines. The powerhouse is 761 feet (232 m) long, 148 feet (45 m) high, 62 feet (19 m) wide and located 1,080 feet (330 m) underground. The two tailrace tunnels are 1691.64 m long. The net head is 312.42 m.

- Cruachan Dam, United Kingdom, built in the early 1960s, a pumped storage plant generating 440 MW from 4 turbines.

- Dinorwig Power Station, Llanberis, United Kingdom, built in 1984, is a pumped-storage system, delivering 1,650 MW to Wales and the north-west of England. It stands in Europe's largest man-made cavern.

- Edward Hyatt Power Plant inside the Oroville Dam, United States, is in a cavern carved into the bedrock of the Feather River canyon. It houses 3 Generator and 3 Pump/Generator units and their respective Transformers 650 feet (200 m) below the crest of the dam.

- Goldisthal Pumped Storage Station, in Thuringia, Germany, built in 1991-2004, generates 1,060 MW from 4 turbines. It is unique (for its scale) in Europe, in that two of the four motor generators are designed as variable speed asynchronous machines. The machine hall is 482 feet (147 m) long, 161 feet (49 m)high, 52 feet (16 m) wide, with a separate transformer cavern (390 feet (120 m) long, 49 feet (15 m)high, 52 feet (16 m) wide).

- Kannagawa Hydropower Plant is under construction in Japan. When completed, it will be the world's largest pumped storage plant, generating 2,700 MW. The power house is 709 feet (216 m) long, 108 feet (33 m) wide, 171 feet (52 m) high. The effective head is 2,343 feet (714 m) The first unit commenced operations in 2005, the second in 2012.

- Kariba hydro-electric power scheme (1,200 MW) is on the Zambezi river, which forms the border between Zimbabwe and Zambia. The Kariba system comprises two underground power stations. The Kariba South station in Zimbabwe houses six 100 MW generators. The Kariba North station in Zambia houses four 150 MW generators.

- Kazunogawa Power Station is a 1,200 MW underground pumped storage plant in Japan. Kazunogawa consists of four 400 MW generation units. The cavern for the underground power station is 1,600 feet (500 m) below the surface. It is 690 feet (210 m) long by 177 feet (54 m) high and 112 feet (34 m) wide. The head is 2,343 feet (714 m).

- Manapouri Power Station, Fiordland, New Zealand, built 1963-1972, generates 850 MW from 7 turbines. It is built 660 feet (200 m) underground, and has two 10 km tailrace tunnels. The net head is 560 feet (170 m). The most notable feature of this station is that the lake and power station are located on the eastern side of the Southern Alps, with the tailrace tunnels traveling under a major mountain range, discharging in Doubtful Sound on the west coast.

- Paulo Afonso Hydroelectric Complex, Brazil; The combined 4,279.6 Paulo Afonso I, II and III were built underground. Completed in 1955, PA I was Brazil's first underground power station.

- Poatina Hydroelectric Power Station, Tasmania, Australia, built in 1966-1977 it generates 300 MW with water provided by the Great Lakes, it is the largest underground power station in Australia.

- Raccoon Mountain Pumped-Storage Plant, Chattanooga, Tennessee, United States, built in 1970-1978 generates 1,530 MW. It is an early test of the pumped-storage approach.

- Robert-Bourassa generating station, Quebec, Canada is the largest underground power station in the world. It generates 5,616 MW from 16 turbines with a net rated head of 450 feet (137.2 m).

Energy Storage

Energy storage is the capture of energy produced at one time for use at a later time. A device that stores energy is sometimes called an accumulator. Energy comes in multiple forms including radiation, chemical, gravitational potential, electrical potential, electricity, elevated temperature, latent heat and kinetic. Energy storage involves converting energy from forms that are difficult to store to more conveniently or economically storable forms. Bulk energy storage is dominated by pumped hydro, which accounts for 99% of global energy storage.

The Llyn Stwlan dam of the Ffestiniog Pumped Storage Scheme in Wales. The lower power station has four water turbines which can generate a total of 360 MW of electricity for several hours, an example of artificial energy storage and conversion.

Some technologies provide short-term energy storage, while others can endure for much longer.

A wind-up clock stores potential energy (in this case mechanical, in the spring tension), a rechargeable battery stores readily convertible chemical energy to operate a mobile phone, and a hydroelectric dam stores energy in a reservoir as gravitational potential energy. Fossil fuels such as coal and gasoline store ancient energy derived from sunlight by organisms that later died, became buried and over time were then converted into these fuels. Food (which is made by the same process as fossil fuels) is a form of energy stored in chemical form.

Ice storage tanks store ice frozen by cheaper energy at night to meet peak daytime demand for cooling. The energy isn't stored directly, but the work-product of consuming energy (pumping away heat) is stored, having the equivalent effect on daytime consumption.

History

Prehistory

The energy present at the initial formation of the universe is stored in stars such as the Sun, and is used by humans directly (e.g. through solar heating or sun tanning), or indirectly (e.g. by growing crops, consuming photosynthesized plants or conversion into electricity in solar cells).

As a purposeful activity, energy storage has existed since pre-history, though it was often not explicitly recognized as such. An example of mechanical energy storage is the use of logs or boulders as defensive measures in ancient forts—the logs or boulders were collected at the top of a hill or wall, and the energy thus stored was used to attack invaders who came within range. Certainly, storing dried wood or another source for fire, or preserving edible food or seeds, in dry or cool areas such as in a cave, under rocks or underground, serves as other examples of energy storage.

Recent History

In the twentieth century grid electrical power was largely generated by burning fossil fuel. When less power was required, less fuel was burned. Concerns with air pollution, energy imports and global warming have spawned the growth of renewable energy such as solar and wind power. Wind power is uncontrolled and may be generating at a time when no additional power is needed. Solar power varies with cloud cover and at best is only available during daylight hours, while demand often peaks after sunset. Interest in storing power from these intermittent sources grows as the renewable energy industry begins to generate a larger fraction of overall energy consumption.

Off grid electrical use was a niche market in the twentieth century, but in the twenty first century it has expanded. Portable devices are in use all over the world. Solar panels are now a common sight in the rural settings worldwide. Access to electricity is now a question of economics, not location. Powering transportation without burning fuel, however, remains in development.

Methods

Outline

The following list includes natural and other non-commercial types of energy storage. in addition to those designed for use in industry and commerce:

- Mechanical
    - Compressed air energy storage (CAES)
    - Fireless locomotive
    - Flywheel energy storage
    - Gravitational potential energy (device)
    - Hydraulic accumulator
    - Liquid nitrogen
    - Pumped-storage hydroelectricity
- Electrical
    - Capacitor
    - Superconducting magnetic energy storage (SMES)
- Biological
    - Glycogen
    - Starch
- Electrochemical
    - Flow battery
    - Rechargeable battery
    - Supercapacitor

- UltraBattery
- Thermal
  - Brick storage heater
  - Cryogenic liquid air or nitrogen
  - Eutectic system
  - Ice storage air conditioning
  - Molten salt
  - Phase Change Material
  - Seasonal thermal energy storage
  - Solar pond
  - Steam accumulator
  - Thermal energy storage (general)
- Chemical
  - Biofuels
  - Hydrated salts
  - Hydrogen
  - Hydrogen peroxide
  - Power to gas
  - Vanadium pentoxide

Mechanical Storage

Energy can be stored in water pumped to a higher elevation using pumped storage methods and also by moving solid matter to higher locations. Other commercial mechanical methods include compressing air and flywheels that convert electric energy into kinetic energy and then back again when electrical demand peaks.

Hydroelectricity

Hydroelectric dams with reservoirs can be operated to provide peak generation at times of peak demand. Water is stored in the reservoir during periods of low demand and released when demand is high. The net effect is similar to pumped storage, but without the pumping loss.

While a hydroelectric dam does not directly store energy from other generating units, it behaves equivalently by lowering output in periods of excess electricity from other sources. In this mode, dams are one of the most efficient forms of energy storage, because only the timing of its generation changes. Hydroelectric turbines have a start-up time on the order of a few minutes.

Pumped-storage

The Sir Adam Beck Generating Complex at Niagara Falls, Canada, which includes a large pumped storage hydroelectricity reservoir to provide an extra 174 MW of electricity during periods of peak demand.

Worldwide, pumped-storage hydroelectricity (PSH) is the largest-capacity form of active grid energy storage available, and, as of March 2012, the Electric Power Research Institute (EPRI) reports that PSH accounts for more than 99% of bulk storage capacity worldwide, representing around 127,000 MW. PSH reported energy efficiency varies in practice between 70% and 80%, with claims of up to 87%.

At times of low electrical demand, excess generation capacity is used to pump water from a lower source into a higher reservoir. When demand grows, water is released back into a lower reservoir (or waterway or body of water) through a turbine, generating electricity. Reversible turbine-generator assemblies act as both a pump and turbine (usually a Francis turbine design). Nearly all facilities use the height difference between two water bodies. Pure pumped-storage plants shift the water between reservoirs, while the "pump-back" approach is a combination of pumped storage and conventional hydroelectric plants that use natural stream-flow.

Compressed Air

A compressed air locomotive used inside a mine between 1928 and 1961.

Compressed air energy storage (CAES) uses surplus energy to compress air for subsequent electricity generation. Small scale systems have long been used in such applications as propulsion of mine locomotives. The compressed air is stored in an underground reservoir.

Compression of air creates heat; the air is warmer after compression. Expansion requires heat. If no extra heat is added, the air will be much colder after expansion. If the heat generated during compression can be stored and used during expansion, efficiency improves considerably. A CAES system can deal with the heat in three ways. Air storage can be adiabatic, diabatic, or isothermal. Another approach uses compressed air to power vehicles.

Flywheel Energy Storage

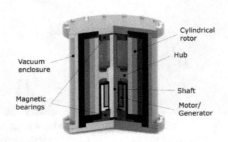

The main components of a typical flywheel.

Flywheel energy storage (FES) works by accelerating a rotor (flywheel) to a very high speed, holding energy as rotational energy. When energy is extracted, the flywheel's rotational speed declines as a consequence of conservation of energy; adding energy correspondingly results in an increase in the speed of the flywheel.

Most FES systems use electricity to accelerate and decelerate the flywheel, but devices that directly use mechanical energy are under consideration.

FES systems have rotors made of high strength carbon-fiber composites, suspended by magnetic bearings and spinning at speeds from 20,000 to over 50,000 rpm in a vacuum enclosure. Such flywheels can reach maximum speed ("charge") in a matter of minutes. The flywheel system is connected to a combination electric motor/generator.

A Flybrid Kinetic Energy Recovery System flywheel. Built for use on Formula 1 racing cars, it is employed to recover and reuse kinetic energy captured during braking.

FES systems have relatively long lifetimes (lasting decades with little or no maintenance; full-cycle lifetimes quoted for flywheels range from in excess of 10^5, up to 10^7, cycles of use), high energy density (100–130 W·h/kg, or 360–500 kJ/kg) and power density.

Gravitational Potential Energy Storage with Solid Masses

Changing the altitude of solid masses can store or release energy via an elevating system driven by an electric motor/generator.

Companies such as Energy Cache and Advanced Rail Energy Storage (ARES) are working on this. ARES uses rails to move concrete weights up and down. Stratosolar proposes to use winches supported by buoyant platforms at an altitude of 20 kilometers, to raise and lower solid masses. Sink Float Solutions proposes to use winches supported by an ocean barge for taking advantage of a 4 km (13,000 ft) elevation difference between the surface and the seabed. ARES estimated a capital cost for the storage capacity of around 60% of pump storage hydroelectricity, Stratosolar $100/kWh and Sink Float Solutions $25/kWh (4000 m depth) and $50/kWh (with 2000 m depth).

Potential energy storage or gravity energy storage was under active development in 2013 in association with the California Independent System Operator. It examined the movement of earth-filled hopper rail cars driven by electric locomotives) from lower to higher elevations.

ARES claimed advantages including indefinite storage with no energy losses, low costs when earth/rocks are used and conservation of water resources.

Thermal Storage

District heating accumulation tower from Theiss near Krems an der Donau in Lower Austria with a thermal capacity of 2 GWh

Thermal storage is the temporary storage or removal of heat. TES is practical because of water's large heat of fusion: the melting of one metric ton of ice (approximately one cubic metre in size) can capture 334 megajoules [MJ] (317,000 BTU) of thermal energy.

An example is Alberta, Canada's Drake Landing Solar Community, for which 97% of the year-round heat is provided by solar-thermal collectors on the garage roofs, with a borehole thermal energy store (BTES) being the enabling technology. STES projects often have paybacks in the four-to-six year range. In Braestrup, Denmark, the community's solar district heating system also utilizes STES, at a storage temperature of 65 °C (149 °F). A heat pump, which is run only when there is surplus wind power available on the national grid, is used to raise the temperature to 80 °C (176 °F) for distribution. When surplus wind generated electricity is not available, a gas-fired boiler is used. Twenty percent of Braestrup's heat is solar.

Latent Heat Thermal Energy Storage (LHTES)

Latent heat thermal energy storage systems works with materials with high latent heat (heat of fusion) capacity, known as phase change materials (PCMs). The main advantage of these materials is that their latent heat storage capacity is much more than sensible heat. In a specific temperature range, phase changes from solid to liquid absorbs a large amount of thermal energy for later use.

Electrochemical
Rechargeable Battery

A rechargeable battery bank used as an uninterruptible power supply in a data center

A rechargeable battery, comprises one or more electrochemical cells. It is known as a 'secondary cell' because its electrochemical reactions are electrically reversible. Rechargeable batteries come in many different shapes and sizes, ranging from button cells to megawatt grid systems.

Rechargeable batteries have lower total cost of use and environmental impact than non-rechargeable (disposable) batteries. Some rechargeable battery types are available in the same form factors as disposables. Rechargeable batteries have higher initial cost but can be recharged very cheaply and used many times.

Common rechargeable battery chemistries include:

- Lead–acid battery: Lead acid batteries hold the largest market share of electric storage products. A single cell produces about 2V when charged. In the charged state the metallic lead negative electrode and the lead sulfate positive electrode are immersed in a dilute sulfuric acid (H_2SO_4) electrolyte. In the discharge process electrons are pushed out of the cell as lead sulfate is formed at the negative electrode while the electrolyte is reduced to water.

- Nickel–cadmium battery (NiCd): Uses nickel oxide hydroxide and metallic cadmium as electrodes. Cadmium is a toxic element, and was banned for most uses by the European Union in 2004. Nickel–cadmium batteries have been almost completely replaced by nickel–metal hydride (NiMH) batteries.

- Nickel–metal hydride battery (NiMH): First commercial types were available in 1989. These are now a common consumer and industrial type. The battery has a hydrogen-absorbing alloy for the negative electrode instead of cadmium.

- Lithium-ion battery: The choice in many consumer electronics and have one of the best energy-to-mass ratios and a very slow self-discharge when not in use.

- Lithium-ion polymer battery: These batteries are light in weight and can be made in any shape desired.

Flow Battery

A flow battery operates by passing a solution over a membrane where ions are exchanged to charge/discharge the cell. Cell voltage is chemically determined by the Nernst equation and ranges, in practical applications, from 1.0 to 2.2 V. Its storage capacity is a function of the volume of the tanks holding the solution.

A flow battery is technically akin both to a fuel cell and an electrochemical accumulator cell. Commercial applications are for long half-cycle storage such as backup grid power.

Supercapacitor

One of a fleet of electric capabuses powered by supercapacitors, at a quick-charge station-bus stop, in service during Expo 2010 Shanghai China. Charging rails can be seen suspended over the bus.

Supercapacitors, also called electric double-layer capacitors (EDLC) or ultracapacitors, are generic terms for a family of electrochemical capacitors that do not have conventional solid dielectrics. Capacitance is determined by two storage principles, double-layer capacitance and pseudocapacitance.

Supercapacitors bridge the gap between conventional capacitors and rechargeable batteries. They store the most energy per unit volume or mass (energy density) among capacitors. They support up to 10,000 farads/1.2 volt, up to 10,000 times that of electrolytic capacitors, but deliver or accept less than half as much power per unit time (power density).

While supercapacitors have energy densities that are approximately 10% of batteries, their power density is generally 10 to 100 times greater. This results in much shorter charge/discharge cycles. Additionally, they will tolerate many more charge and discharge cycles than batteries.

Supercapacitors support a broad spectrum of applications, including:

- Low supply current for memory backup in static random-access memory (SRAM)
- Power for cars, buses, trains, cranes and elevators, including energy recovery from braking, short-term energy storage and burst-mode power delivery

Ultra Battery

The UltraBattery is a hybrid lead-acid cell and carbon-based ultracapacitor (or supercapacitor) invented by Australia's national research body, the Commonwealth Scientific and Industrial Research Organisation (CSIRO). The lead-acid cell and ultracapacitor share the sulfuric acid electrolyte and both are packaged into the same physical unit. The UltraBattery can be manufactured with similar physical and electrical characteristics to conventional lead-acid batteries making it possible to cost-effectively replace many lead-acid applications.

The UltraBattery tolerates high charge and discharge levels and endures large numbers of cycles, outperforming previous lead-acid cells by more than an order of magnitude. In hybrid-electric vehicle tests, millions of cycles have been achieved. The UltraBattery is also highly tolerant to the effects of sulfation compared with traditional lead-acid cells. This means it can operate continuously in partial state of charge whereas traditional lead-acid batteries are generally held at full charge between discharge events. It is generally electrically inefficient to fully charge a lead-acid battery so by decreasing time spent in the top region of charge the UltraBattery achieves high efficiencies, typically between 85 and 95% DC-DC.

The UltraBattery can work across a wide range of applications. The constant cycling and fast charging and discharging necessary for applications such as grid regulation and leveling and electric vehicles can damage chemical batteries, but are well handled by the ultracapacitive qualities of UltraBattery technology. The technology has been installed in Australia and the US on the megawatt scale, performing frequency regulation and renewable smoothing applications.

Other Chemical

Power to Gas

Power to gas is a technology which converts electricity into a gaseous fuel such as hydrogen or methane. The three commercial methods use electricity to reduce water into hydrogen and oxygen by means of electrolysis.

In the first method, hydrogen is injected into the natural gas grid or is used in transport or industry. The second method is to combine the hydrogen with carbon dioxide to produce methane using a methanation reaction such as the Sabatier reaction, or biological methanation, resulting in an extra energy conversion loss of 8%. The methane may then be fed into the natural gas grid. The third method uses the output gas of a wood gas generator or a biogas plant, after the biogas upgrader is mixed with the hydrogen from the electrolyzer, to upgrade the quality of the biogas.

Hydrogen

The element hydrogen can be a form of stored energy. Hydrogen can produce electricity via a hydrogen fuel cell.

At penetrations below 20% of the grid demand, renewables do not severely change the economics; but beyond about 20% of the total demand, external storage becomes important. If these sources are used to make ionic hydrogen, they can be freely expanded. A 5-year community-based pilot program using wind turbines and hydrogen generators began in 2007 in the remote community of Ramea, Newfoundland and Labrador. A similar project began in 2004 on Utsira, a small Norwegian island.

Energy losses involved in the hydrogen storage cycle come from the electrolysis of water, liquification or compression of the hydrogen and conversion to electricity.

About 50 kW·h (180 MJ) of solar energy is required to produce a kilogram of hydrogen, so the cost of the electricity is crucial. At $0.03/kWh, a common off-peak high-voltage line rate in the United States, hydrogen costs $1.50 a kilogram for the electricity, equivalent to $1.50/gallon for gasoline. Other costs include the electrolyzer plant, hydrogen compressors or liquefaction, storage and transportation.

Underground hydrogen storage is the practice of hydrogen storage in underground caverns, salt domes and depleted oil and gas fields. Large quantities of gaseous hydrogen have been stored in underground caverns by Imperial Chemical Industries for many years without any difficulties. The European Hyunder project indicated in 2013 that storage of wind and solar energy using underground hydrogen would require 85 caverns.

Methane

Methane is the simplest hydrocarbon with the molecular formula CH_4. Methane is more easily stored than hydrogen and the transportation. Storage and combustion infrastructure (pipelines, gasometers, power plants) are mature.

Synthetic natural gas (syngas or SNG) can be created in a multi-step process, starting with hydrogen and oxygen. Hydrogen is then reacted with carbon dioxide in a Sabatier process, producing methane and water. Methane can be stored and later used to produce electricity. The resulting water is recycled, reducing the need for water. In the electrolysis stage oxygen is stored for methane combustion in a pure oxygen environment at an adjacent power plant, eliminating nitrogen oxides.

Methane combustion produces carbon dioxide (CO_2) and water. The carbon dioxide can be recycled to boost the Sabatier process and water can be recycled for further electrolysis. Methane production, storage and combustion recycles the reaction products.

The CO_2 has economic value as a component of an energy storage vector, not a cost as in carbon capture and storage.

Power to Liquid

Power to liquid is similar to power to gas, however the hydrogen produced by electrolysis from wind and solar electricity isn't converted into gases such as methane but into liquids such as methanol. Methanol is easier in handling than gases and requires less safety precautions than hydrogen. It can be used for transportation, including aircraft, but also for industrial purposes or in the power sector.

Biofuels

Various biofuels such as biodiesel, vegetable oil, alcohol fuels, or biomass can replace fossil fuels. Various chemical processes can convert the carbon and hydrogen in coal, natural gas, plant and animal biomass and organic wastes into short hydrocarbons suitable as replacements for existing hydrocarbon fuels. Examples are Fischer–Tropsch diesel, methanol, dimethyl ether and syngas. This diesel source was used extensively in World War II in Germany, which faced limited access to crude oil supplies. South Africa produces most of the country's diesel from coal for similar reasons. A long term oil price above US$35/bbl may make such large scale synthetic liquid fuels economical.

Aluminium, Boron, Silicon, and Zinc

Aluminium, Boron, silicon, lithium, and zinc have been proposed as energy storage solutions.

Electrical Methods

Capacitor

This mylar-film, oil-filled capacitor has very low inductance and low resistance, to provide the high-power (70 megawatts) and the very high speed (1.2 microsecond) discharges needed to operate a dye laser.

A capacitor (originally known as a 'condenser') is a passive two-terminal electrical component used to store energy electrostatically. Practical capacitors vary widely, but all contain at least two electrical conductors (plates) separated by a dielectric (i.e., insulator). A capacitor can store electric energy when disconnected from its charging circuit, so it can be used like a temporary battery, or like other types of rechargeable energy storage system. Capacitors are commonly used in electronic devices to maintain power supply while batteries change. (This prevents loss of information in volatile memory.) Conventional capacitors provide less than 360 joules per kilogram, while a conventional alkaline battery has a density of 590 kJ/kg.

Capacitors store energy in an electrostatic field between their plates. Given a potential difference across the conductors (e.g., when a capacitor is attached across a battery), an electric field develops across the dielectric, causing positive charge (+Q) to collect on one plate and negative charge (-Q) to collect on the other plate. If a battery is attached to a capacitor for a sufficient amount of time, no current can flow through the capacitor. However, if an accelerating or alternating voltage is applied across the leads of the capacitor, a displacement current can flow.

Capacitance is greater given a narrower separation between conductors and when the conductors have a larger surface area. In practice, the dielectric between the plates emits a small amount of leakage current and has an electric field strength limit, known as the breakdown voltage. The conductors and leads introduce undesired inductance and resistance.

Research is assessing the quantum effects of nanoscale capacitors for digital quantum batteries.

Superconducting Magnetics

Superconducting magnetic energy storage (SMES) systems store energy in a magnetic field created by the flow of direct current in a superconducting coil that has been cooled to a temperature below its superconducting critical temperature. A typical SMES system includes a superconducting coil, power conditioning system and refrigerator. Once the superconducting coil is charged, the current does not decay and the magnetic energy can be stored indefinitely.

The stored energy can be released to the network by discharging the coil. The associated inverter/rectifier accounts for about 2–3% energy loss in each direction. SMES loses the least amount of electricity in the energy storage process compared to other methods of storing energy. SMES systems offer round-trip efficiency greater than 95%.

Due to the energy requirements of refrigeration and the cost of superconducting wire, SMES is used for short duration storage such as improving power quality. It also has applications in grid balancing.

Interseasonal Thermal Storage

Seasonal thermal energy storage (STES) allows heat or cold to be used months after it was collected from waste energy or natural sources. The material can be stored in contained aquifers, clusters of boreholes in geological substrates such as sand or crystalline bedrock, in lined pits filled with gravel and water, or water-filled mines.

Applications

Mills

A more recent application is the control of waterways to drive water mills for processing grain or powering machinery. Complex systems of reservoirs and dams were constructed to store and release water (and the potential energy it contained) when required.

Home Energy Storage

Home energy storage is expected to become increasingly present given the growing importance of distributed generation (especially photovoltaics) and the important share of energy consumption in buildings. A household equipped with photovoltaics can achieve a maximum electricity self-sufficiency of about 40%. To reach higher levels of self-sufficiency, energy storage is needed, given the mismatch between energy consumption and energy production from photovoltaics. In 2015, multiple manufacturers announced rechargeable battery systems for storing energy, generally to hold surplus energy from home solar/wind generation.

Tesla Motors announced the first two models of the Tesla Powerwall. One is a 10 kWh weekly cycle version for backup applications and the other is a 7 kWh version for daily cycle applications.

Enphase Energy announced an integrated system that allows home users to store, monitor and manage electricity. The system stores 1.2 kWh hours of energy and 275W/500W power output.

Grid Electricity

Renewable Energy Storage

Construction of the Salt Tanks which provide efficient thermal energy storage so that output can be provided after the sun goes down, and output can be scheduled to meet demand requirements. The 280 MW Solana Generating Station is designed to provide six hours of energy storage. This allows the plant to generate about 38 percent of its rated capacity over the course of a year.

The 150 MW Andasol solar power station is a commercial parabolic trough solar thermal power plant, located in Spain. The Andasol plant uses tanks of molten salt to store solar energy so that it can continue generating electricity even when the sun isn't shining.

The largest source and the greatest store of renewable energy is provided by hydroelectric dams. A large reservoir behind a dam can store enough water to average the annual flow of a river between dry and wet seasons. A very large reservoir can store enough water to average the flow of a river between dry and wet years. While a hydroelectric dam does not directly store energy from intermittent sources, it does balance the grid by lowering its output and retaining its water when power is generated by solar or wind. If wind or solar generation exceeds the regions hydroelectric capacity, then some additional source of energy will be needed.

Many renewable energy sources (notably solar and wind) produce variable power. Storage systems can level out the imbalances between supply and demand that this causes. Electricity must be used as it is generated or converted immediately into storable forms.

The main method of electrical grid storage is pumped-storage hydroelectricity. Areas of the world such as Norway, Wales, Japan and the US have used elevated geographic features for reservoirs, using electrically powered pumps to fill them. When needed, the water passes through generators and converts the gravitational potential of the falling water into electricity. Pumped storage in Norway, which gets almost all its electricity from hydro, has an instantaneous capacity of 25–30 GW expandable to 60 GW—enough to be "Europe's battery".

Some forms of storage that produce electricity include pumped-storage hydroelectric dams, rechargeable batteries, thermal storage including molten salts which can efficiently store and release very large quantities of heat energy, and compressed air energy storage, flywheels, cryogenic systems and superconducting magnetic coils.

Surplus power can also be converted into methane (sabatier process) with stockage in the natural gas network.

In 2011, the Bonneville Power Administration in Northwestern United States created an experimental program to absorb excess wind and hydro power generated at night or during stormy periods that are accompanied by high winds. Under central control, home appliances absorb surplus energy by heating ceramic bricks in special space heaters to hundreds of degrees and by boosting the temperature of modified hot water heater tanks. After charging, the appliances provide home heating and hot water as needed. The experimental system was created as a result of a severe 2010 storm that overproduced renewable energy to the extent that all conventional power sources were shut down, or in the case of a nuclear power plant, reduced to its lowest possible operating level, leaving a large area running almost completely on renewable energy.

Another advanced method used at the Solar Project in the United States and the Solar Tres Power Tower in Spain uses molten salt to store thermal energy captured from the sun and then convert it and dispatch it as electrical power. The system pumps molten salt through a tower or other special conduits to be heated by the sun. Insulated tanks store the solution. Electricity is produced by turning water to steam that is fed to turbines.

Since the early 21st century batteries have been applied to utility scale load-leveling and frequency regulation capabilities.

In vehicle-to-grid storage, electric vehicles that are plugged into the energy grid can deliver stored electrical energy from their batteries into the grid when needed.

Generation

Chemical fuels remain the dominant form of energy storage for electricity generation. Natural gas is crowding out other forms, such as oil and coal.

Air Conditioning

Thermal energy storage (TES) can be used for air conditioning. It is most widely used for cooling single large buildings and/or groups of smaller buildings. Commercial air conditioning systems are the biggest contributors to peak electrical loads. In 2009, thermal storage was used in over 3,300 buildings in over 35 countries. It works by creating ice at night and using the ice to for cooling during the hotter daytime periods.

The most popular technique is ice storage, which requires less space than water and is less costly than fuel cells or flywheels. In this application, a standard chiller runs at night to produce an ice pile. Water then circulates through the pile during the day to chill water that would normally be the chiller's daytime output.

A partial storage system minimizes capital investment by running the chillers nearly 24 hours a day. At night, they produce ice for storage and during the day they chill water. Water circulating through the melting ice augments the production of chilled water. Such a system makes ice for 16 to 18 hours a day and melts ice for six hours a day. Capital expenditures are reduced because the chillers can be just 40 - 50% of the size needed for a conventional, no-storage design. Storage sufficient to store half a day's available heat is usually adequate.

A full storage system shuts off the chillers during peak load hours. Capital costs are higher, as such a system requires larger chillers and a larger ice storage system.

This ice is produced when electrical utility rates are lower. Off-peak cooling systems can lower energy costs. The U.S. Green Building Council has developed the Leadership in Energy and Environmental Design (LEED) program to encourage the design of reduced-environmental impact buildings. Off-peak cooling may help toward LEED Certification.

Thermal storage for heating is less common than for cooling. An example of thermal storage is storing solar heat to be used for heating at night.

Latent heat can also be stored in technical phase change materials (PCMs). These can be encapsulated in wall and ceiling panels, to moderate room temperatures.

Transport

Liquid hydrocarbon fuels are the most commonly used forms of energy storage for use in transportation. Other energy carriers such as hydrogen can be used to avoid producing greenhouse gases.

Electronics

Capacitors are widely used in electronic circuits for blocking direct current while allowing alternating current to pass. In analog filter networks, they smooth the output of power supplies. In resonant circuits they tune radios to particular frequencies. In electric power transmission systems they stabilize voltage and power flow.

Use Cases

The United States Department of Energy International Energy Storage Database (IESDB), is a free-access database of energy storage projects and policies funded by the United States Department of Energy Office of Electricity and Sandia National Labs.

Economics

The economics of Energy Storage strictly depends on the reserve service requested, and several uncertainty factors affect the profitability of Energy Storage. Therefore not every Energy Storage is technically and economically suitable for the storage of several MWh, and the optimal size of the Energy Storage is market and location dependent.

Moreover, ESS are affected by several risks, e.g.:

1) techno-economic risks, which are related to the specific technology;

2) Market risks, which are the factors that affect the electricity supply system;

3) Regulation and policy risks.

Therefore, traditional techniques based on deterministic Discounted Cash Flow (DCF) for the investment appraisal are not fully adequate to evaluate these risks and uncertainties and the investor's flexibility to deal with them. Hence, the literature recommends to assess the value of risks and uncertainties through the Real Option Analysis (ROA), which is a valuable method in uncertain contexts.

The economic valuation of large-scale applications (including pumped hydro storage and compressed air) considers benefits including: wind curtailment avoidance, grid congestion avoidance, price arbitrage and carbon free energy delivery. In one technical assessment by the Carnegie Mellon Electricity Industry Centre, economic goals could be met with batteries if energy storage were achievable at a capital cost of $30 to $50 per kilowatt-hour of storage capacity.

A metric for calculating the energy efficiency of storage systems is Energy Storage On Energy Invested (ESOI) which is the useful energy used to make the storage system divided into the lifetime energy storage. For lithium ion batteries this is around 10, and for lead acid batteries it is about 2. Other forms of storage such as pumped hydroelectric storage generally have higher ESOI, such as 210.

Research

Germany

The German Federal government has allocated €200M (approximately US$270M) for advanced research, as well as providing a further €50M to subsidize battery storage for use with residential rooftop solar panels, according to a representative of the German Energy Storage Association.

Siemens AG commissioned a production-research plant to open in 2015 at the *Zentrum für Sonnenenergie und Wasserstoff (ZSW*, the German Center for Solar Energy and Hydrogen Research in the State of Baden-Württemberg), a university/industry collaboration in Stuttgart, Ulm and Widderstall, staffed by approximately 350 scientists, researchers, engineers, and technicians. The plant develops new near-production manufacturing materials and processes (NPMM&P) using a computerized Supervisory Control and Data Acquisition (SCADA) system. Its goals will enable the expansion of rechargeable battery production with both increased quality and reduced manufacturing costs.

United States

In 2014, research and test centers opened to evaluate energy storage technologies. Among them was the Advanced Systems Test Laboratory at the University of Wisconsin at Madison in Wisconsin State, which partnered with battery manufacturer Johnson Controls. The laboratory was created as part of the university's newly opened Wisconsin Energy Institute. Their goals include the evaluation of state-of-the-art and next generation electric vehicle batteries, including their use as grid supplements.

The State of New York unveiled its New York Battery and Energy Storage Technology (NY-BEST) Test and Commercialization Center at Eastman Business Park in Rochester, New York, at a cost of $23 million for its almost 1,700 m² laboratory. The center includes the Center for Future Energy Systems, a collaboration between Cornell University of Ithaca, New York and the Rensselaer Polytechnic Institute in Troy, New York. NY-BEST tests, validates and independently certifies diverse forms of energy storage intended for commercial use.

United Kingdom

In the United Kingdom, some fourteen industry and government agencies allied with seven British universities in May 2014 to create the SUPERGEN Energy Storage Hub in order to assist in the coordination of energy storage technology research and development.

Pumped-storage Hydroelectricity

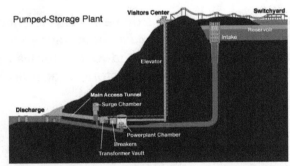

Diagram of the TVA pumped storage facility at Raccoon Mountain Pumped-Storage Plant

Shaded-relief topo map of the Taum Sauk pumped storage plant in Missouri

Pumped-storage hydroelectricity (PSH), or pumped hydroelectric energy storage (PHES), is a type of hydroelectric energy storage used by electric power systems for load balancing. The method stores energy in the form of gravitational potential energy of water, pumped from a lower elevation reservoir to a higher elevation. Low-cost off-peak electric power is used to run the pumps. During periods of high electrical demand, the stored water is released through turbines to produce electric power. Although the losses of the pumping process makes the plant a net consumer of energy overall, the system increases revenue by selling more electricity during periods of *peak demand*, when electricity prices are highest. Pumped-storage hydroelectricity allows energy from intermittent sources (such as solar, wind) and other renewables, or excess electricity from continuous baseload sources (such as coal or nuclear) to be saved for periods of higher demand.

Pumped storage is the largest-capacity form of grid energy storage available, and, as of March 2012, the Electric Power Research Institute (EPRI) reports that PSH accounts for more than 99% of bulk production capacity worldwide, representing around 127 GW, with storage capacity at 740 TWh. Typically, the *round-trip* energy efficiency of PSH varies in practice between 70% and 80%, with some claiming up to 87%. The main disadvantage of PHS is the specialist nature of the site required, needing both geographical height and water availability. Suitable sites are therefore likely to be in hilly or mountainous regions, and potentially in areas of outstanding natural beauty, and therefore there are also social and ecological issues to overcome.

Overview

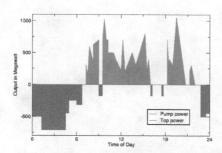

Power distribution, over a day, of a pumped-storage hydroelectricity facility. Green represents power consumed in pumping; red is power generated.

At times of low electrical demand, excess generation capacity is used to pump water into the higher reservoir. When there is higher demand, water is released back into the lower reservoir through a turbine, generating electricity. Reversible turbine/generator assemblies act as pump and turbine (usually a Francis turbine design). Nearly all facilities use the height difference between two natural bodies of water or artificial reservoirs. Pure pumped-storage plants just shift the water between reservoirs, while the "pump-back" approach is a combination of pumped storage and conventional hydroelectric plants that use natural stream-flow. Plants that do not use pumped-storage are referred to as conventional hydroelectric plants; conventional hydroelectric plants that have significant storage capacity may be able to play a similar role in the electrical grid as pumped storage, by deferring output until needed.

Taking into account evaporation losses from the exposed water surface and conversion losses, energy recovery of 80% or more can be regained. The technique is currently the most cost-effective means of storing large amounts of electrical energy on an operating basis, but capital costs and the presence of appropriate geography are critical decision factors.

The relatively low energy density of pumped storage systems requires either large flows or a large differences in height. The only way to store a significant amount of energy is by having a large body of water located relatively near, but as high above as possible, a second body of water. In some places this occurs naturally, in others one or both bodies of water were man-made. Projects in which both reservoirs are artificial and in which no natural waterways are involved are commonly referred to as "closed loop".

This system may be economical because it flattens out load variations on the power grid, permitting thermal power stations such as coal-fired plants and nuclear power plants that provide baseload electricity to continue operating at peak efficiency (Base load power plants), while reducing the need for "peaking" power plants that use the same fuels as many base-load thermal plants, gas and oil, but have been designed for flexibility rather than maximal thermal efficiency. Hence pumped storage systems are crucial when coordinating large groups of heterogeneous generators. However, capital costs for purpose-built hydrostorage are relatively high.

Along with energy management, pumped storage systems help control electrical network frequency and provide reserve generation. Thermal plants are much less able to respond to sudden changes in electrical demand, potentially causing frequency and voltage instability. Pumped storage plants, like other hydroelectric plants, can respond to load changes within seconds.

The upper reservoir (Llyn Stwlan) and dam of the Ffestiniog Pumped Storage Scheme in north Wales. The lower power station has four water turbines which generate 360 MW of electricity within 60 seconds of the need arising.

The first use of pumped storage was in the 1890s in Italy and Switzerland. In the 1930s reversible hydroelectric turbines became available. These turbines could operate as both turbine-generators and in reverse as electric motor driven pumps. The latest in large-scale engineering technology are variable speed machines for greater efficiency. These machines generate in synchronization with the network frequency, but operate asynchronously (independent of the network frequency) as motor-pumps.

The first use of pumped-storage in the United States was in 1930 by the Connecticut Electric and Power Company, using a large reservoir located near New Milford, Connecticut, pumping water from the Housatonic River to the storage reservoir 230 feet above.

The important use for pumped storage is to level the fluctuating output of intermittent energy sources. The pumped storage provides a load at times of high electricity output and low electricity demand, enabling additional system peak capacity. In certain jurisdictions, electricity prices may be close to zero or occasionally negative (Ontario in early September, 2006), on occasions that there is more electrical generation than load available to absorb it; although at present this is rarely due to wind alone, increased wind generation may increase the likelihood of such occurrences. It is particularly likely that pumped storage will become especially important as a balance for very large scale photovoltaic generation.

Worldwide Use

In 2009, world pumped storage generating capacity was 104 GW, while other sources claim 127 GW, which comprises the vast majority of all types of utility grade electric storage. The EU had 38.3 GW net capacity (36.8% of world capacity) out of a total of 140 GW of hydropower and representing 5% of total net electrical capacity in the EU. Japan had 25.5 GW net capacity (24.5% of world capacity).

In 2010 the United States had 21.5 GW of pumped storage generating capacity (20.6% of world capacity). PHS generated (net) -5.501 GWh of energy in 2010 in the US because more energy is consumed in pumping than is generated.

The five largest operational pumped-storage plants are listed below:

Station	Country	Location	Capacity (MW)
Bath County Pumped Storage Station	United States	38°12′32″N 79°48′00″W38.20889°N 79.80000°W	3,003
Guangdong Pumped Storage Power Station	China	23°45′52″N 113°57′12″E23.76444°N 113.95333°E	2,400
Huizhou Pumped Storage Power Station	China	23°16′07″N 114°18′50″E23.26861°N 114.31389°E	2,400
Okutataragi Pumped Storage Power Station	Japan	35°14′13″N 134°49′55″E35.23694°N 134.83194°E	1,932
Ludington Pumped Storage Power Plant	United States	43°53′37″N 86°26′43″W43.89361°N 86.44528°W	1,872
Note: this table shows the power-generating capacity in megawatts as is usual for power stations. However, the overall energy-storage capacity in megawatt-hours (MWh) is a different intrinsic property and can not be derived from the above given figures.			

Potential Technologies

Sea Water

One can use pumped sea water to store the energy. In 1999, the 30 MW Yanbaru project in Okinawa was the first demonstration of seawater pumped storage. A 300 MW seawater-based project has recently been proposed on Lanai, Hawaii, and several seawater-based projects have recently been proposed in Ireland and Chile. Another potential example of this could be used in a tidal barrage or tidal lagoon. A potential benefit of this arises if seawater is allowed to flow behind the barrage or into the lagoon at high tide when the water level is roughly equal either side of the barrier, when the potential energy difference is close to zero. Then water is released at low tide when a head of water has been built up behind the barrier, when there is a far greater potential energy difference between the two bodies of water. The result being that when the energy used to pump the water is recovered, it will have multiplied to a degree depending on the head of water built up. A further enhancement is to pump more water at high tide further increasing the head with for example intermittent renewables. Two downsides are that the generator must be below sea level, and that marine organisms would tend to grow on the equipment and disrupt operation. This is not a major problem for the EDF La Rance Tidal power station in France.

Direct Pumping

A new concept is to use wind turbines or solar power to drive water pumps directly, in effect an 'Energy Storing Wind or Solar Dam'. This could provide a more efficient process and usefully smooth out the variability of energy captured from the wind or sun. . In northern Chile, the Espejo de Tarapacá project is a power project of Valhalla, that combines solar and hydroelectric resources. The project takes advantage of the unique geographic characteristics of the Atacama Desert in order to build a 300 MW pumped storage hydroelectric plant that uses the Pacific Ocean as its lower reservoir and an existing natural concavity as its upper reservoir, and a 600 MW-AC solar photovoltaic plant that is located in the region with the highest solar irradiation in the world. The above characteristics minimize the environmental impact and the cost of the plant, which is USD 400 MM for the pumped storage and USD 900 MM for the solar power plant.

Underground Reservoirs

The use of underground reservoirs has been investigated. Recent examples include the proposed Summit project in Norton, Ohio, the proposed Maysville project in Kentucky (underground limestone mine), and the Mount Hope project in New Jersey, which was to have used a former iron mine as the lower reservoir. Several new underground pumped storage projects have been proposed. Cost-per-kilowatt estimates for these projects can be lower than for surface projects if they use existing underground mine space. There are limited opportunities involving suitable underground space, but the number of underground pumped storage opportunities may increase if abandoned coal mines prove suitable.

Decentralised Systems

Small pumped-storage hydropower plants can be built on streams and within infrastructures, such as drinking water networks and artificial snow making infrastructures. Such plants provide distributed energy storage and distributed flexible electricity production and can contribute to the decentralized integration of intermittent renewable energy technologies, such as wind power and Solar power. Reservoirs that can be used for small pumped-storage hydropower plants could include natural or artificial lakes, reservoirs within other structures such as irrigation, or unused portions of mines or underground military installations. In Switzerland one study suggested that the total installed capacity of small pumped-storage hydropower plants in 2011 could be increased by 3 to 9 times by providing adequate policy instruments.

References

- Bent Sørensen (2004). Renewable Energy: Its Physics, Engineering, Use, Environmental Impacts, Economy, and Planning Aspects. Academic Press. pp. 556–. ISBN 978-0-12-656153-1.

- Dwivedi, A.K. Raja, Amit Prakash Srivastava, Manish (2006). Power Plant Engineering. New Delhi: New Age International. p. 354. ISBN 81-224-1831-7.

- Raghunath, H.M. (2009). Hydrology : principles, analysis, and design (Rev. 2nd ed.). New Delhi: New Age International. p. 288. ISBN 81-224-1825-2.

- McNeil, Ian (1996). An Encyclopaedia of the History of Technology ([New ed.]. ed.). London: Routledge. p. 369. ISBN 978-0-415-14792-7.

- Wiser, Wendell H. (2000). Energy resources: occurrence, production, conversion, use. Birkhäuser. p. 190. ISBN 978-0-387-98744-6.

- British Electricity International (1991). Modern Power Station Practice: incorporating modern power system practice (3rd Edition (12 volume set) ed.). Pergamon. ISBN 0-08-040510-X.

- Thomas C. Elliott, Kao Chen, Robert Swanekamp (coauthors) (1997). Standard Handbook of Powerplant Engineering (2nd ed.). McGraw-Hill Professional. ISBN 0-07-019435-1.

- Dandekar, M. M.; Sharma, K. N. (2010). Water power engineering. Noida: Vikas Publishing House. p. 381. ISBN 0706986369. Retrieved 25 January 2015.

- Aifantis, Katerina E.; Hackney, Stephen A.; Kumar, R. Vasant (March 30, 2010). High Energy Density Lithium Batteries: Materials, Engineering, Applications. John Wiley & Sons. ISBN 978-3-527-63002-8.

- B. E. Conway (1999). Electrochemical Supercapacitors: Scientific Fundamentals and Technological Applications. Berlin: Springer. ISBN 0306457369. Retrieved May 2, 2013.

- Guilherme de Oliveira e Silva; Patrick Hendrick (September 15, 2016). "Lead-acid batteries coupled with photovoltaics for increased electricity self-sufficiency in households". Retrieved July 20, 2016.

- "UNIDO, ICSHP Launch Small Hydropower Knowledge Sharing Portal". Sustainable Energy Policy and Practice. Retrieved 29 April 2015.

- "2013 Accomplishment Report" (PDF). Small Power Utilities Group, National Power Corporation. Retrieved 15 September 2015.

- Arthur Williams. "The Performance of Centrifugal Pumps as Turbines and Influence of Pump Geometry" (PDF). Esha.be. Retrieved 2015-08-12.

- CraiginPA (19 February 2015). "Cerrowire 50 ft. 18-Gauge 2 Conductor Thermostat Wire-210-1002B - The Home Depot". The Home Depot. Retrieved 8 August 2015.

- Donnie (29 May 2015). "Southwire 4 Stranded THHN Black (By-the-Foot)-20499099 - The Home Depot". The Home Depot. Retrieved 8 August 2015.

- A K Raja, Amit Prakash Shriwastava, Manish Dwivedi. Power Plant Engineering. Digital Designs. pp. 358–359. Retrieved 25 January 2015.

- Debord, Matthew (May 1, 2015). "Elon Musk's big announcement: it's called 'Tesla Energy'". Business Insider. Retrieved June 11, 2015.

- Delacey, Lynda (October 29, 2015). "Enphase plug-and-play solar energy storage system to begin pilot program". www.gizmag.com. Retrieved December 20, 2015.

Tidal Power: Generation Process and Devices

Tidal power is a form of hydropower; it helps in the conversion of energy that is obtained from tides to useful forms such as electricity. The features explained in the section are tidal stream generator, tidal barrage, SeaGen and shrouded tidal turbine. The major components of tidal power are discussed in this section.

Tidal Power

Tidal power, also called tidal energy, is a form of hydropower that converts the energy obtained from tides into useful forms of power, mainly electricity.

Although not yet widely used, tidal power has potential for future electricity generation. Tides are more predictable than wind energy and solar power. Among sources of renewable energy, tidal power has traditionally suffered from relatively high cost and limited availability of sites with sufficiently high tidal ranges or flow velocities, thus constricting its total availability. However, many recent technological developments and improvements, both in design (e.g. dynamic tidal power, tidal lagoons) and turbine technology (e.g. new axial turbines, cross flow turbines), indicate that the total availability of tidal power may be much higher than previously assumed, and that economic and environmental costs may be brought down to competitive levels.

Historically, tide mills have been used both in Europe and on the Atlantic coast of North America. The incoming water was contained in large storage ponds, and as the tide went out, it turned waterwheels that used the mechanical power it produced to mill grain. The earliest occurrences date from the Middle Ages, or even from Roman times. It was only in the 19th century that the process of using falling water and spinning turbines to create electricity was introduced in the U.S. and Europe.

The world's first large-scale tidal power plant was the Rance Tidal Power Station in France, which became operational in 1966. It was the largest tidal power station in terms of output until Sihwa Lake Tidal Power Station opened in South Korea in August, 2011. The Sihwa station uses sea wall defense barriers complete with 10 turbines generating 254 MW.

Total harvestable energy from tidal areas close to a coast is estimated to be around 1 terawatt worldwide.

Generation of Tidal Energy

Tidal power is taken from the Earth's oceanic tides. Tidal forces are periodic variations in gravitational attraction exerted by celestial bodies. These forces create corresponding motions or currents in the world's oceans. Due to the strong attraction to the oceans, a bulge in the water level is created, causing a temporary increase in sea level. When the sea level is raised, water from the middle of the ocean is forced to move toward the shorelines, creating a tide. This occurrence takes place in an unfailing manner, due to the consistent pattern of the moon's orbit around the earth. The magnitude and character of this motion reflects the changing positions of the Moon and Sun relative to the Earth, the effects of Earth's rotation, and local geography of the sea floor and coastlines.

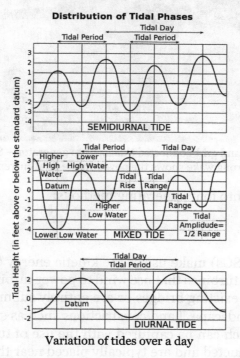

Variation of tides over a day

Tidal power is the only technology that draws on energy inherent in the orbital characteristics of the Earth–Moon system, and to a lesser extent in the Earth–Sun system. Other natural energies exploited by human technology originate directly or indirectly with the Sun, including fossil fuel, conventional hydroelectric, wind, biofuel, wave and solar energy. Nuclear energy makes use of Earth's mineral deposits of fissionable elements, while geothermal power taps the Earth's internal heat, which comes from a combination of residual heat from planetary accretion (about 20%) and heat produced through radioactive decay (80%).

A tidal generator converts the energy of tidal flows into electricity. Greater tidal variation and higher tidal current velocities can dramatically increase the potential of a site for tidal electricity generation.

Because the Earth's tides are ultimately due to gravitational interaction with the Moon and Sun and the Earth's rotation, tidal power is practically inexhaustible and classified as a renewable energy resource. Movement of tides causes a loss of mechanical energy in the Earth–Moon system: this is a result of pumping of water through natural restrictions around coastlines and consequent viscous dissipation at the seabed and in turbulence. This loss of energy has caused the rotation of the Earth to slow in the 4.5 billion years since its formation. During the last 620 million years the period of rotation of the earth (length of a day) has increased from 21.9 hours to 24 hours; in this period the Earth has lost 17% of its rotational energy. While tidal power will take additional energy from the system, the effect is negligible and would only be noticed over millions of years.

Generating Methods

The world's first commercial-scale and grid-connected tidal stream generator – SeaGen – in Strangford Lough. The strong wake shows the power in the tidal current.

Tidal power can be classified into four generating methods:

Tidal Stream Generator

Tidal stream generators (or TSGs) make use of the kinetic energy of moving water to power turbines, in a similar way to wind turbines that use wind to power turbines. Some tidal generators can be built into the structures of existing bridges or are entirely submersed, thus avoiding concerns over impact on the natural landscape. Land constrictions such as straits or inlets can create high velocities at specific sites, which can be captured with the use of turbines. These turbines can be horizontal, vertical, open, or ducted and are typically placed near the bottom of the water column where tidal velocities are greatest.

Tidal Barrage

Tidal barrages make use of the potential energy in the difference in height (or hydraulic head) between high and low tides. When using tidal barrages to generate power, the potential energy from a tide is seized through strategic placement of specialized dams. When the sea level rises and the tide begins to come in, the temporary increase in tidal power is channeled into a large basin behind the dam, holding a large amount of potential energy. With the receding tide, this energy is then converted into mechanical energy as the water is released through large turbines that create electrical power through the use of generators. Barrages are essentially dams across the full width of a tidal estuary.

Dynamic Tidal Power

Dynamic tidal power (or DTP) is an untried but promising technology that would exploit an interaction between potential and kinetic energies in tidal flows. It proposes that very long dams (for example: 30–50 km length) be built from coasts straight out into the sea or ocean, without enclosing an area. Tidal phase differences are introduced across the dam, leading to a significant water-level differential in shallow coastal seas – featuring strong coast-parallel oscillating tidal currents such as found in the UK, China, and Korea.

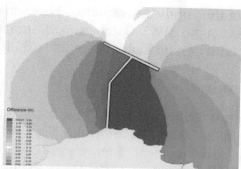

Top-down view of a DTP dam. Blue and dark red colors indicate low and high tides, respectively.

Tidal Lagoon

A newer tidal energy design option is to construct circular retaining walls embedded with turbines that can capture the potential energy of tides. The created reservoirs are similar to those of tidal barrages, except that the location is artificial and does not contain a preexisting ecosystem. The lagoons can also be in double (or triple) format without pumping or with pumping that will flatten out the power output. The pumping power could be provided by excess to grid demand renewable energy from for example wind turbines or solar photovoltaic arrays. Excess renewable energy rather than being curtailed could be used and stored for a later period of time. Geographically dispersed tidal lagoons with a time delay between peak production would also flatten out peak production providing near base load production though at a higher cost than some other alternatives such as district heating renewable energy storage. The proposed Tidal Lagoon Swansea Bay in Wales, United Kingdom would be the first tidal power station of this type once built.

US and Canadian Studies in the Twentieth Century

The first study of large scale tidal power plants was by the US Federal Power Commission in 1924 which if built would have been located in the northern border area of the US state of Maine and the south eastern border area of the Canadian province of New Brunswick, with various dams, power-houses, and ship locks enclosing the Bay of Fundy and Passamaquoddy Bay. Nothing came of the study and it is unknown whether Canada had been approached about the study by the US Federal Power Commission.

In 1956, utility Nova Scotia Light and Power of Halifax commissioned a pair of studies into the feasibility of commercial tidal power development on the Nova Scotia side of the Bay of Fundy. The two studies, by Stone & Webster of Boston and by Montreal Engineering Company of Montreal independently concluded that millions of horsepower could be harnessed from Fundy but that development costs would be commercially prohibitive at that time.

There was also a report on the international commission in April 1961 entitled "Investigation of the International Passamaquoddy Tidal Power Project" produced by both the US and Canadian Federal Governments. According to benefit to costs ratios, the project was beneficial to the US but not to Canada. A highway system along the top of the dams was envisioned as well.

A study was commissioned by the Canadian, Nova Scotian and New Brunswick governments (Reassessment of Fundy Tidal Power) to determine the potential for tidal barrages at Chignecto Bay and Minas Basin – at the end of the Fundy Bay estuary. There were three sites determined to be financially feasible: Shepody Bay (1550 MW), Cumberline Basin (1085 MW), and Cobequid Bay (3800 MW). These were never built despite their apparent feasibility in 1977.

Tidal Power Development in the UK

The world's first marine energy test facility was established in 2003 to start the development of the wave and tidal energy industry in the UK. Based in Orkney, Scotland, the European Marine Energy Centre (EMEC) has supported the deployment of more wave and tidal energy devices than at any other single site in the world. EMEC provides a variety of test sites in real sea conditions. Its grid connected tidal test site is located at the Fall of Warness, off the island of Eday, in a narrow channel which concentrates the tide as it flows between the Atlantic Ocean and North Sea. This area has a very strong tidal current, which can travel up to 4 m/s (8 knots) in spring tides. Tidal energy developers that have tested at the site include: Alstom (formerly Tidal Generation Ltd); ANDRITZ HYDRO Hammerfest; Atlantis Resources Corporation; Nautricity; OpenHydro; Scotrenewables Tidal Power; Voith. The resource could be 4 TJ per year.

Current and Future Tidal Power Schemes

- The first tidal power station was the Rance tidal power plant built over a period of 6 years from 1960 to 1966 at La Rance, France. It has 240 MW installed capacity.

- 254 MW Sihwa Lake Tidal Power Plant in South Korea is the largest tidal power installation in the world. Construction was completed in 2011.

- The first tidal power site in North America is the Annapolis Royal Generating Station, Annapolis Royal, Nova Scotia, which opened in 1984 on an inlet of the Bay of Fundy. It has 20 MW installed capacity.

- The Jiangxia Tidal Power Station, south of Hangzhou in China has been operational since 1985, with current installed capacity of 3.2 MW. More tidal power is planned near the mouth of the Yalu River.

- The first in-stream tidal current generator in North America (Race Rocks Tidal Power Demonstration Project) was installed at Race Rocks on southern Vancouver Island in September 2006. The next phase in the development of this tidal current generator will be in Nova Scotia (Bay of Fundy).

- A small project was built by the Soviet Union at Kislaya Guba on the Barents Sea. It has 0.4 MW installed capacity. In 2006 it was upgraded with a 1.2MW experimental advanced orthogonal turbine.

- Jindo Uldolmok Tidal Power Plant in South Korea is a tidal stream generation scheme planned to be expanded progressively to 90 MW of capacity by 2013. The first 1 MW was installed in May 2009.

- A 1.2 MW SeaGen system became operational in late 2008 on Strangford Lough in Northern Ireland.

- The contract for an 812 MW tidal barrage near Ganghwa Island (South Korea) north-west of Incheon has been signed by Daewoo. Completion is planned for 2015.

- A 1,320 MW barrage built around islands west of Incheon is proposed by the South Korean government, with projected construction starting in 2017.

- The Scottish Government has approved plans for a 10MW array of tidal stream generators near Islay, Scotland, costing 40 million pounds, and consisting of 10 turbines – enough to power over 5,000 homes. The first turbine is expected to be in operation by 2013.

- The Indian state of Gujarat is planning to host South Asia's first commercial-scale tidal power station. The company Atlantis Resources planned to install a 50MW tidal farm in the Gulf of Kutch on India's west coast, with construction starting early in 2012.

- Ocean Renewable Power Corporation was the first company to deliver tidal power to the US grid in September, 2012 when its pilot TidGen system was successfully deployed in Cobscook Bay, near Eastport.

- In New York City, 30 tidal turbines will be installed by Verdant Power in the East River by 2015 with a capacity of 1.05MW.

- Construction of a 320 MW tidal lagoon power plant outside the city of Swansea in the UK was granted planning permission in June 2015 and work is expected to start in 2016. Once completed, it will generate over 500GWh of electricity per year, enough to power roughly 155,000 homes.

- A turbine project is being installed in Ramsey Sound in 2014.

- The largest tidal energy project entitled "Meygen" (398MW) is currently in construction in the Pentland Firth in northern Scotland

Tidal Power Issues

Environmental Concerns

Tidal power can have effects on marine life. The turbines can accidentally kill swimming sea life with the rotating blades, although projects such as the one in Strangford feature a safety mechanism that turns off the turbine when marine animals approach. Some fish may no longer utilize the area if threatened with a constant rotating or noise-making object. Marine life is a huge factor when placing tidal power energy generators in the water and precautions are made to ensure that as many marine animals as possible will not be affected by it. The Tethys database provides access to scientific literature and general information on the potential environmental effects of tidal energy.

Tidal Turbines

The main environmental concern with tidal energy is associated with blade strike and entanglement of marine organisms as high speed water increases the risk of organisms being pushed near or through these devices. As with all offshore renewable energies, there is also a concern about how the creation of EMF and acoustic outputs may affect marine organisms. It should be noted that because these devices are in the water, the acoustic output can be greater than those created with offshore wind energy. Depending on the frequency and amplitude of sound generated by the tidal energy devices, this acoustic output can have varying effects on marine mammals (particularly those who echolocate to communicate and navigate in the marine environment, such as dolphins and whales). Tidal energy removal can also cause environmental concerns such as degrading far-field water quality and disrupting sediment processes. Depending on the size of the project, these effects can range from small traces of sediment building up near the tidal device to severely affecting nearshore ecosystems and processes.

Tidal Barrage

Installing a barrage may change the shoreline within the bay or estuary, affecting a large ecosystem that depends on tidal flats. Inhibiting the flow of water in and out of the bay, there may also be less flushing of the bay or estuary, causing additional turbidity (suspended solids) and less salt-water, which may result in the death of fish that act as a vital food source to birds and mammals. Migrating fish may also be unable to access breeding streams, and may attempt to pass through the turbines. The same acoustic concerns apply to tidal barrages. Decreasing shipping accessibility can become a socio-economic issue, though locks can be added to allow slow passage. However, the barrage may improve the local economy by increasing land access as a bridge. Calmer waters may also allow better recreation in the bay or estuary. In August 2004, a humpback whale swam through the open sluice gate of the Annapolis Royal Generating Station at slack tide, ending up trapped for several days before eventually finding its way out to the Annapolis Basin.

Tidal Lagoon

Environmentally, the main concerns are blade strike on fish attempting to enter the lagoon, acoustic output from turbines, and changes in sedimentation processes. However, all these effects are localized and do not affect the entire estuary or bay.

Corrosion

Salt water causes corrosion in metal parts. It can be difficult to maintain tidal stream generators due to their size and depth in the water. The use of corrosion-resistant materials such as stainless steels, high-nickel alloys, copper-nickel alloys, nickel-copper alloys and titanium can greatly reduce, or eliminate, corrosion damage.

Mechanical fluids, such as lubricants, can leak out, which may be harmful to the marine life nearby. Proper maintenance can minimize the amount of harmful chemicals that may enter the environment.

Fouling

The biological events that happen when placing any structure in an area of high tidal currents and high biological productivity in the ocean will ensure that the structure becomes an ideal substrate for the growth of marine organisms. In the references of the Tidal Current Project at Race Rocks in British Columbia this is documented. Also see this page and Several structural materials and coatings were tested by the Lester Pearson College divers to assist Clean Current in reducing fouling on the turbine and other underwater infrastructure.

Structural Health Monitoring

The high load factors resulting from the fact that water is 800 times denser than air and the predictable and reliable nature of tides compared with the wind makes tidal energy particularly attractive for Electric power generation. Condition monitoring is the key for exploiting it cost-efficiently.

Dynamic Tidal Power

Co-inventor Kees Hulsbergen presenting the principles of DTP at Tsinghua University in Beijing, in February 2010.

Dynamic tidal power or DTP is an untried but promising technology for tidal power generation. It would involve creating a long dam-like structure perpendicular to the coast, with the option for a coast-parallel barrier at the far end, forming a large 'T' shape. This long T-dam would interfere with coast-parallel tidal wave hydrodynamics, creating water level differences on opposite sides of the barrier which drive a series of bi-directional turbines installed in the dam. Oscillating tidal waves which run along the coasts of continental shelves, containing powerful hydraulic currents, are common in *e.g.* China, Korea, and the UK.

The concept was invented and patented in 1997 by Dutch coastal engineers Kees Hulsbergen and Rob Steijn.

A short video explaining the concept was completed in October 2013 and made available in English on YouTube and in Chinese on Youku.

Description

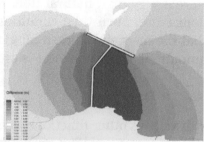

Top-down view of a DTP dam. Blue and dark red colors indicate low and high tides, respectively.

A DTP dam is a long barrier of 30 km or more which is built perpendicular to the coast, running straight out into the sea, without enclosing an area. Along many coasts of the world, the main tidal movement runs parallel to the coastline: the entire mass of the ocean water accelerates in one direction, and later in the day back the other way. A DTP dam is long enough to exert an influence on the horizontal tidal movement, which generates a water level differential (head) over both sides of the dam. The head can be converted into power, using a long series of conventional low-head turbines installed in the dam.

Maximum Head Difference

Estimates of the maximum head difference that can be obtained from a variety of dam configurations are based on numerical and analytical models. Field information from measured water level differences across natural barriers confirms the creation of significant head. The (maximum) head difference is more than what would be expected in stationary flow situations (such as rivers). The maximum head difference reaches values up to a few m, which can be attributed to the non-permanent character of the tidal flow (acceleration).

Benefits

High Power Output

It is estimated that some of the largest dams could accommodate over 15 GW (15000 MW) of installed capacity. A DTP dam with 8 GW installed capacity and a capacity factor of about 30%, could generate about 21 TWh annually. To put this number in perspective, an average European person consumes about 6800 kWh per year, so one DTP dam could supply energy for about 3 million Europeans.

Stable Power

The generation of tidal power is highly predictable due to the deterministic nature of tides, and independent of weather conditions or climate change. Power output varies with the tidal phase (ebb & flow, neap & spring) but the shorter terms effects can be avoided by combining two dams, placed at certain distance from each other (in the order of 150–250 km), each generating maximum electricity output when the other is generating minimal output. This provides a predictable and fairly stable base generation to the energy grid.

High Availability

Dynamic tidal power doesn't require a very high natural tidal range, but instead an open coast where the tidal propagation is alongshore. Such tidal conditions can be found in many places around the world, which means that the theoretical potential of DTP is very high. Along the Chinese coast for example, the total amount of available power is estimated at 80 - 150 GW.

Potential for Combined Functions

The long dam can be combined with various other functions, such as coastal protection, deep sea – and LNG ports, aquaculture facilities, controlled land reclamation and connections between islands and the mainland. These additional functions can share the investment costs, thus helping to lower the price per kWh.

Challenges

A major challenge is that the proof of DTP functioning can only be demonstrated by putting it in practice. Testing the concept of DTP at a small scale within a demonstration project, would not be effective, since almost no power would be yielded. Not even at a dam length of 1 km or so, because the DTP principle is such that the power generation capacity increases as the square of the dam length increases (both head and volume increase in a more or less linear manner for increased dam length, resulting in a quadratic increase in power generation). Economic viability is estimated to be reached for dam lengths of about 30 km.

Demonstration Project

A demonstration project under consideration in China would not involve construction of a dam, but instead feature a newly cut channel through a long peninsula with a narrow isthmus (neck). The channel would feature a head of about 1 – 2 meters, and be fitted with low-head bi-directional turbines, similar to the type which would be used for full-scale DTP.

Status of Technological Development

No DTP dam has ever been built, although all of the technologies required to build a DTP dam are available. Various mathematical and physical models have been conducted to model and predict the 'head' or water level differential over a dynamic tidal power dam. The interaction between tides and long dams has been observed and recorded in large engineering projects, such as the Delta Works and the Afsluitdijk in the Netherlands. The interaction of tidal currents with natural peninsulas is also well-known, and such data is used to calibrate numerical models of tides. Formulas for the calculation of added mass were applied to develop an analytical model of DTP. Observed water level differentials closely match current analytical and numerical models. Water level differential generated over a DTP dam can now be predicted with a useful degree of accuracy.

Some of the key elements required include:

- Bi-directional turbines (capable of generating power in both directions) for low head, high-volume environments. Operational units exist for seawater applications, reaching an efficiency of over 75%.

- Dam construction methods. This could be achieved by modular floating caissons (concrete building blocks). These caissons would be manufactured on shore and subsequently floated to the dam location.

- Suitable sites to demonstrate DTP. A pilot project of DTP could be integrated with a planned coastal development project, such as a sea bridge, island connection, deep sea port, land reclamation, offshore wind farm, etc., built in a suitable environment for DTP.

Recent Progress

In December 2011 the Dutch Ministry of Economy, Agriculture and Innovation (EL&I) awarded a grant funding subsidy to the POWER consortium, led by Strukton and managed by ARCADIS. The maximum grant is about 930.000 euro, which is matched by a similar amount of co-financing from the consortium partners. The POWER group conducts a detailed feasibility study on the development of Dynamic Tidal Power (DTP) in China in a 3-year programme jointly conducted with Chinese government institutes. The commitments of the programme to achieve by 2015, registered under the UN Sustainable Energy for All initiative include:

- Determine most suitable sites for DTP implementation in China, Korea, and the UK

- Complete detailed feasibility studies for two DTP pilot power plants in China

- Complete pre-feasibility study for one full-scale DTP power plant in China

- Worldwide dissemination of technical information regarding DTP among relevant target groups

In August 2012, China's National Energy Administration formed a consortium of companies and research institutes, led by the Hydropower and Water Resources Planning and Design General Institute (also known as China Renewable Energy Engineering Institute), to investigate DTP. A bilateral agreement on DTP cooperation was signed between China and the Netherlands on September 27, 2012. Following technical exchange to verify the principles, a modelling study was conducted to select sites. In October 2013, a more in-depth economic analysis study was started to better understand the economic costs and benefits of DTP.

A short video explaining the concept was completed in October 2013 and made available in English on YouTube and in Chinese on Youku.

Tidal Stream Generator

Evopod - A semi-submerged floating approach tested in Strangford Lough.

A tidal stream generator, often referred to as a tidal energy converter (TEC) is a machine that extracts energy from moving masses of water, in particular tides, although the term is often used in reference to machines designed to extract energy from run of river or tidal estuarine sites. Certain types of these machines function very much like underwater wind turbines, and are thus often referred to as tidal turbines. They were first conceived in the 1970s during the oil crisis.

Tidal stream generators are the cheapest and the least ecologically damaging among the three main forms of tidal power generation.

Similarity to Wind Turbines

Tidal stream generators draw energy from water currents in much the same way as wind turbines draw energy from air currents. However, the potential for power generation by an individual tidal turbine can be greater than that of similarly rated wind energy turbine. The higher density of water relative to air (water is about 800 times the density of air) means that a single generator can provide significant power at low tidal flow velocities compared with similar wind speed. Given that power varies with the density of medium and the cube of velocity, water speeds of nearly one-tenth the speed of wind provide the same power for the same size of turbine system; however this limits the application in practice to places where tide speed is at least 2 knots (1 m/s) even close to neap tides. Furthermore, at higher speeds in a flow between 2 and 3 metres per second in seawater a tidal turbine can typically access four times as much energy per rotor swept area as a similarly rated power wind turbine.

Types of Tidal Stream Generators

No standard tidal stream generator has emerged as the clear winner, among a large variety of designs. Several prototypes have shown promise with many companies making bold claims, some of which are yet to be independently verified, but they have not operated commercially for extended periods to establish performances and rates of return on investments.

The European Marine Energy Centre recognizes six principal types of tidal energy converter. They are horizontal axis turbines, vertical axis turbines, oscillating hydrofoils, venturi devices, Archimedes screws and tidal kites.

Axial Turbines

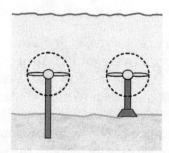

Bottom-mounted axial turbines

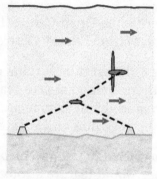

A cable tethered turbine

These are close in concept to traditional windmills, but operating under the sea. They have most of the prototypes currently operating, including:

Tocardo, a Dutch-based company, has been running tidal turbines since 2008 on the Afsluitdijk, near Den Oever. Typical production data of tidal generator shown of the T100 model as applied in Den Oever. Currently 1 River model (R1) and 2 tidal models (T) are in production with a 3rd T3 coming soon. Power production for the T1 is around 100 kW and around 200 kW for the T2.

The AR-1000, a 1MW turbine developed by Atlantis Resources Corporation that was successfully deployed at the EMEC facility during the summer of 2011. The AR series are commercial scale, horizontal axis turbines designed for open ocean deployment. AR turbines feature a single rotor set with fixed pitch blades. The AR turbine is rotated as required with each tidal exchange. This is done in the slack period between tides and held in place for the optimal heading for the next tide. AR turbines are rated at 1MW @ 2.65 m/s of water flow velocity.

The Kvalsund installation is south of Hammerfest, Norway. Although still a prototype, a turbine with a reported capacity of 300 kW was connected to the grid on 13 November 2003.

Seaflow, a 300 kW Periodflow marine current propeller type turbine was installed by Marine Current Turbines off the coast of Lynmouth, Devon, England, in 2003. The 11m diameter turbine generator was fitted to a steel pile which was driven into the seabed. As a prototype, it was connected to a dump load, not to the grid.

In April 2007 Verdant Power began running a prototype project in the East River between Queens and Roosevelt Island in New York City; it was the first major tidal-power project in the United States. The strong currents pose challenges to the design: the blades of the 2006 and 2007 prototypes broke and new reinforced turbines were installed in September 2008.

Following the Seaflow trial, a full-size prototype, called SeaGen, was installed by Marine Current Turbines in Strangford Lough in Northern Ireland in April 2008. The turbine began to generate at full power of just over 1.2 MW in December 2008 and is reported to have fed 150 kW into the grid for the first time on 17 July 2008, and has now contributed more than a gigawatt hour to consumers in Northern Ireland. It is currently the only commercial scale device to have been installed anywhere in the world. SeaGen is made up of two axial flow rotors, each of which drive a generator. The turbines are capable of generating electricity on both the ebb and flood tides because the rotor blades can pitch through 180°.

OpenHydro, an Irish company exploiting the Open-Centre Turbine developed in the U.S., has a prototype being tested at the European Marine Energy Centre (EMEC), in Orkney, Scotland.

A prototype semi-submerged floating tethered tidal turbine called Evopod has been tested since June 2008 in Strangford Lough, Northern Ireland at 1/10 scale. The UK company developing it is called Ocean Flow Energy Ltd. The advanced hull form maintains optimum heading into the tidal stream and is designed to operate in the peak flow of the water column.

In 2010, Tenax Energy of Australia proposed to put 450 turbines off the coast of Darwin, Australia, in the Clarence Strait. The turbines would feature a rotor section approximately 15 metres in diameter with a slightly larger gravity base. The turbines would operate in deep water well below shipping channels. Each turbine is forecast to produce energy for between 300 and 400 homes.

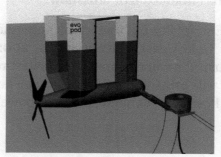

A 3D model of an Evopod tidal turbine

Tidalstream, a UK-based company, commissioned a scaled-down Triton 3 turbine in the Thames. It can be floated to its site, installed without cranes, jack-ups or divers and then ballasted into operating position. At full scale the Triton 3 in 30-50m deep water has a 3MW capacity, and the Triton 6 in 60-80m water has a capacity of up to 10MW, depending on the flow. Both platforms have man-access capability both in the operating position and in the float-out maintenance position.

Crossflow Turbines

Invented by Georges Darreius in 1923 and patented in 1929, these turbines can be deployed either vertically or horizontally.

The Gorlov turbine is a variant of the Darrieus design featuring a helical design that is in a large scale, commercial pilot in South Korea, starting with a 1MW plant that opened in May 2009 and expanding to 90MW by 2013. Neptune Renewable Energy's Proteus project employs a shrouded vertical axis turbine that can be used to form an array in mainly estuarine conditions.

In April 2008, the Ocean Renewable Power Company, LLC (ORPC) successfully completed testing its proprietary turbine-generator unit (TGU) prototype at ORPC's Cobscook Bay and Western Passage tidal sites near Eastport, Maine. The TGU is the core of the OCGen technology and utilizes advanced design cross-flow (ADCF) turbines to drive a permanent magnet generator located between the turbines and mounted on the same shaft. ORPC has developed TGU designs that can be used for generating power from river, tidal and deep water ocean currents.

Trials in the Strait of Messina, Italy, started in 2001 of the Kobold turbine concept.

Flow Augmented Turbines

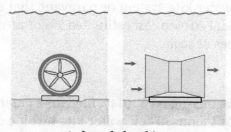

A shrouded turbine

Using flow augmentation measures, for example a duct or shroud, the incident power available to a turbine can be increased. The most common example uses a shroud to increase the flow rate through the turbine, which can be either axial or crossflow.

The Australian company Tidal Energy Pty Ltd undertook successful commercial trials of efficient shrouded tidal turbines on the Gold Coast, Queensland in 2002. Tidal Energy delivered their shrouded turbine in northern Australia where some of the fastest recorded flows (11 m/s, 21 knots) are found. Two small turbines will provide 3.5 MW. Another larger 5 meter diameter turbine, capable of 800 kW in 4 m/s of flow, was planned as a tidal powered desalination showcase near Brisbane Australia.

Oscillating Devices

Oscillating devices do not have a rotating component, instead making use of aerofoil sections which are pushed sideways by the flow. Oscillating stream power extraction was proven with the omni- or bi-directional Wing'd Pump windmill. During 2003 a 150 kW oscillating hydroplane device, the Stingray, was tested off the Scottish coast. The Stingray uses hydrofoils to create oscillation, which allows it to create hydraulic power. This hydraulic power is then used to power a hydraulic motor, which then turns a generator.

Pulse Tidal operate an oscillating hydrofoil device in the Humber estuary. Having secured funding from the EU, they are developing a commercial scale device to be commissioned 2012.

The bioSTREAM tidal power conversion system, uses the biomimicry of swimming species, such as shark, tuna, and mackerel using their highly efficient Thunniform mode propulsion. It is produced by Australian company BioPower Systems.

A 2 kW prototype relying on the use of two oscillating hydrofoils in a tandem configuration has been developed at Laval University and tested successfully near Quebec City, Canada, in 2009. A hydrodynamic efficiency of 40% has been achieved during the field tests.

Venturi Effect

Venturi effect devices use a shroud or duct in order to generate a pressure differential which is used to run a secondary hydraulic circuit which is used to generate power. A device, the Hydro Venturi, is to be tested in San Francisco Bay.

Tidal Kite Turbines

A tidal kite turbine is an underwater kite system or paravane that converts tidal energy into electricity by moving through the tidal stream. An estimated 1% of 2011's global energy requirements could be provided by such devices at scale.

History

Ernst Souczek of Vienna, Austria, on August 6, 1947, filed for a patent US2501696; assignor of one-half to Wolfgang Kmentt, also of Vienna. Their water kite turbine disclosure demonstrated a rich art in water-kite turbines. In similar technology, many others prior to 2006 advanced water-kite and paravane electric generating systems. In 2006, a tidal kite turbine was developed by Swedish company Minesto. They conducted its first sea trial in Strangford Lough in Northern Ireland in the summer of 2011. The test used kites with wingspan of 1.4m. In 2013 the Deep Green pilot plant began operation off Northern Ireland. The plant uses carbon fiber kites with a wingspan of 8m (or 12m). Each kite has a rated power of 120 kilowatts at a tidal flow of 1.3 meters per second.

Design

Minesto's kite has a wingspan of 8–14 metres (26–46 ft). The kite has neutral buoyancy, so doesn't sink as the tide turns from ebb to flow. Each kite is equipped with a gearless turbine to generate which is transmitted by the attachment cable to a transformer and then to the electricity grid. The turbine mouth is protected to protect marine life. The 14-meter version has a rated power of 850 kilowatts at 1.7 meters per second.

Operation

The kite is tethered by a cable to a fixed point. It "flies" through the current carrying a turbine. It moves in a figure-eight loop to increase the speed of the water flowing through the turbine tenfold. Force increases with the cube of velocity, offering the potential to generate 1,000-fold more energy than a stationary generator. That maneuver means the kite can operate in tidal streams that move too slowly to drive earlier tidal devices, such as the SeaGen turbine. The kite was expected to work in flows as low 1–2.5 metres (3 ft 3 in–8 ft 2 in) per second, while first-generation devices need over 2.5s. Each kite will have a capacity to generate between 150 and 800 kW. They can be deployed in waters 50–300 metres (160–980 ft) deep.

Tidal Stream Developers

There are a number of individuals and companies developing tidal energy converters across the world. A database of all know tidal energy developers is kept up-to-date here: Tidal energy developers

Tidal Stream Testing

The world's first marine energy test facility was established in 2003 to kick start the development of the wave and tidal energy industry in the UK. Based in Orkney, Scotland, the European Marine Energy Centre (EMEC) has supported the deployment of more wave and tidal energy devices than at any other single site in the world. EMEC provides a variety of test sites in real sea conditions. It's grid connected tidal test site is located at the Fall of Warness, off the island of Eday, in a narrow channel which concentrates the tide as it flows between the Atlantic Ocean and North Sea. This area has a very strong tidal current, which can travel up to 4 m/s (8 knots) in spring tides. Tidal energy developers currently testing at the site include Alstom (formerly Tidal Generation Ltd), ANDRITZ HYDRO Hammerfest, OpenHydro, Scotrenewables Tidal Power, and Voith.

Commercial Plans

RWE's npower announced that it is in partnership with Marine Current Turbines to build a tidal farm of SeaGen turbines off the coast of Anglesey in Wales, near the Skerries.

"The Skerries project located in Anglesey, Wales, will be one of the first arrays deployed using the Siemens owned Marine Current Turbines SeaGen S tidal turbines. The marine consent for the project was recently awarded, the first tidal array to be consented in Wales. The 10MW array will be fully operational in 2015." - CEO of Siemens Energy Hydro & Ocean Unit Achim Wörner

In November 2007, British company Lunar Energy announced that, in conjunction with E.ON, they would be building the world's first deep-sea tidal energy farm off the coast of Pembrokeshire in Wales. It will provide electricity for 5,000 homes. Eight underwater turbines, each 25 metres long and 15 metres high, are to be installed on the sea bottom off St David's peninsula. Construction is due to start in the summer of 2008 and the proposed tidal energy turbines, described as "a wind farm under the sea", should be operational by 2010.

British Columbia Tidal Energy Corp. plans to deploy at least three 1.2 MW turbines in the Campbell River or in the surrounding coastline of British Columbia by 2009.

Alderney Renewable Energy Ltd was granted a licence in 2008 and is planning to use tidal turbines to extract power from the notoriously strong tidal races around Alderney in the Channel Islands. It is estimated that up to 3 GW could be extracted. This would not only supply the island's needs but also leave a considerable surplus for export, utilising a France-Alderney-Britain cable (FAB Link) which is expected to go online by 2020.

Nova Scotia Power has selected OpenHydro's turbine for a tidal energy demonstration project in the Bay of Fundy, Nova Scotia, Canada and Alderney Renewable Energy Ltd for the supply of tidal turbines in the Channel Islands.

Pulse Tidal are designing a commercial device with seven other companies who are expert in their fields. The consortium was awarded an €8 million EU grant to develop the first device, which will be deployed in 2012 and generate enough power for 1,000 homes.

ScottishPower Renewables are planning to deploy ten 1MW HS1000 devices designed by Hammerfest Strom in the Sound of Islay.

In March 2014, the Federal Energy Regulatory Committee (FERC) approved a pilot license for Snohomish County PUD to install two OpenHydro tidal turbines in Admiralty Inlet, WA. This project is the first grid-connected two-turbine project in the US; installation is planned for the summer of 2015. The OpenHydro tidal turbines that Snohomish County PUD will use are designed to be placed directly into the seafloor at a depth of roughly 200 feet, so that there will be no effect on commercial navigation overhead. The license granted by the FERC also includes plans to protect fish, wildlife, as well as cultural and aesthetic resources, in addition to navigation. Each turbine measures 6 meters in diameter, and will generate up to 300 kW of electricity.

Energy Calculations

Turbine Power

Tidal energy converters can have varying modes of operating and therefore varying power output. If the power coefficient of the device "C_p" is known, the equation below can be used to determine the power output of the hydrodynamic subsystem of the machine. This available power cannot exceed that imposed by the Betz limit on the power coefficient, although this can be circumvented to some degree by placing a turbine in a shroud or duct. This works, in essence, by forcing water which would not have flowed through the turbine through the rotor disk. In these situations it is the frontal area of the duct, rather than the turbine, which is used in calculating the power coefficient and therefore the Betz limit still applies to the device as a whole.

The energy available from these kinetic systems can be expressed as:

$$P = \frac{\rho A V^3}{2} C_P$$

where:

C_P = the turbine power coefficient

P = the power generated (in watts)

ρ = the density of the water (seawater is 1027 kg/m³)

A = the sweep area of the turbine (in m²)

V = the velocity of the flow

Relative to an open turbine in free stream, ducted turbines are capable of as much as 3 to 4 times the power of the same turbine rotor in open flow.

Resource Assessment

While initial assessments of the available energy in a channel have focus on calculations using the kinetic energy flux model, the limitations of tidal power generation are significantly more complicated. For example, the maximum physical possible energy extraction from a strait connecting two large basins is given to within 10% by:

$$P = 0.22 \rho g \Delta H_{max} Q_{max}$$

where

ρ = the density of the water (seawater is 1027 kg/m³)

g = gravitational acceleration (9.80665 m/s²)

ΔH_{max} = maximum differential water surface elevation across the channel

Q_{max} = maximum volumetric flow rate though the channel.

Potential Sites

As with wind power, selection of location is critical for the tidal turbine. Tidal stream systems need to be located in areas with fast currents where natural flows are concentrated between obstructions, for example at the entrances to bays and rivers, around rocky points, headlands, or between islands or other land masses. The following potential sites are under serious consideration:

- Pembrokeshire in Wales
- River Severn between Wales and England
- Cook Strait in New Zealand
- Kaipara Harbour in New Zealand
- Bay of Fundy in Canada.
- East River in the United States
- Golden Gate in the San Francisco Bay
- Piscataqua River in New Hampshire
- The Race of Alderney and The Swinge in the Channel Islands
- The Sound of Islay, between Islay and Jura in Scotland
- Pentland Firth between Caithness and the Orkney Islands, Scotland
- Humboldt County, California in the United States
- Columbia River, Oregon in the United States
- Plaquemines Parish, Louisiana in the Southern United States
- Isle of Wight, England
- Teddington and Ham Hydro at Teddington on the River Thames in the London suburbs, England

Modern advances in turbine technology may eventually see large amounts of power generated from the ocean, especially tidal currents using the tidal stream designs but also from the major thermal current systems such as the Gulf Stream, which is covered by the more general term marine current power. Tidal stream turbines may be arrayed in high-velocity areas where natural tidal current flows are concentrated such as the west and east coasts of Canada, the Strait of Gibraltar, the Bosporus, and numerous sites in Southeast Asia and Australia. Such flows occur almost anywhere where there are entrances to bays and rivers, or between land masses where water currents are concentrated.

Environmental Impacts

The main environmental concern with tidal energy is associated with blade strike and entanglement of marine organisms as high speed water increases the risk of organisms being pushed near or through these devices. As with all offshore renewable energies, there is also a concern about how the creation of EMF and acoustic outputs may affect marine organisms. It should be noted that because these devices are in the water, the acoustic output can be greater than those created with offshore wind energy. Depending on the frequency and amplitude of sound generated by the tidal energy devices, this acoustic output can have varying effects on marine mammals (particularly those who echolocate to communicate and navigate in the marine environment such as dolphins and whales). Tidal energy removal can also cause environmental concerns such as degrading farfield water quality and disrupting sediment processes. Depending on the size of the project, these effects can range from small traces of sediment build up near the tidal device to severely affecting nearshore ecosystems and processes.

One study of the Roosevelt Island Tidal Energy (RITE, Verdant Power) project in the East River (New York City), utilized 24 split beam hydroacoustic sensors (scientific echosounder) to detect and track the movement of fish both upstream and downstream of each of six turbines. The results suggested (1) very few fish using this portion of the river, (2) those fish which did use this area were not using the portion of the river which would subject them to blade strikes, and (3) no evidence of fish traveling through blade areas.

Work is currently being conducted by the Northwest National Marine Renewable Energy Center (NNMREC)to explore and establish tools and protocols for assessment of physical and biological conditions and monitor environmental changes associated with tidal energy development.

Tidal Barrage

The Rance Tidal Power Station, a tidal barrage in France.

A tidal barrage is a dam-like structure used to capture the energy from masses of water moving in and out of a bay or river due to tidal forces.

Instead of damming water on one side like a conventional dam, a tidal barrage first allows water to flow into a bay or river during high tide, and releasing the water back during low tide. This is done by measuring the tidal flow and controlling the sluice gates at key times of the tidal cycle. Turbines are then placed at these sluices to capture the energy as the water flows in and out.

Tidal barrages are among the oldest methods of tidal power generation, with projects being developed as early as the 1960s, such as the 1.7 megawatt Kislaya Guba Tidal Power Station in Kislaya Guba, Russia.

Generating Methods

The barrage method of extracting tidal energy involves building a barrage across a bay or river that is subject to tidal flow. Turbines installed in the barrage wall generate power as water flows in and out of the estuary basin, bay, or river. These systems are similar to a hydro dam that produces static head or pressure head (a height of water pressure). When the water level outside of the basin or lagoon changes relative to the water level inside, the turbines are able to produce power.

An artistic impression of a tidal barrage, including embankments, a ship lock and caissons housing a sluice and two turbines.

The basic elements of a barrage are caissons, embankments, sluices, turbines, and ship locks. Sluices, turbines, and ship locks are housed in caissons (very large concrete blocks). Embankments seal a basin where it is not sealed by caissons.

The sluice gates applicable to tidal power are the flap gate, vertical rising gate, radial gate, and rising sector.

Only a few such plants exist. The first was the Rance Tidal Power Station, on the Rance river, in France, which has been operating since 1966, and generates 240MW. A larger 254MW plant began operation at Sihwa Lake, Korea, in 2011. Smaller plants include one on the Bay of Fundy, and another across a tiny inlet in Kislaya Guba, Russia. A number of proposals have been considered for a Severn barrage across the River Severn, from Brean Down in England to Lavernock Point near Cardiff in Wales.

Barrage systems are affected by problems of high civil infrastructure costs associated with what is in effect a dam being placed across estuarine systems, and the environmental problems associated with changing a large ecosystem.

Ebb Generation

The basin is filled through the sluices until high tide. Then the sluice gates are closed. (At this stage there may be "Pumping" to raise the level further). The turbine gates are kept closed until the sea level falls to create sufficient head across the barrage, and then are opened so that the turbines generate until the head is again low. Then the sluices are opened, turbines disconnected and the basin is filled again. The cycle repeats itself. Ebb generation (also known as outflow generation) takes its name because generation occurs as the tide changes tidal direction.

Flood Generation

The basin is filled through the turbines, which generate at tide flood. This is generally much less efficient than ebb generation, because the volume contained in the upper half of the basin (which is where ebb generation operates) is greater than the volume of the lower half (filled first during flood generation). Therefore, the available level difference – important for the turbine power produced – between the basin side and the sea side of the barrage, reduces more quickly than it would in ebb generation. Rivers flowing into the basin may further reduce the energy potential, instead of enhancing it as in ebb generation. Of course this is not a problem with the "lagoon" model, without river inflow.

Pumping

Turbines are able to be powered in reverse by excess energy in the grid to increase the water level in the basin at high tide (for ebb generation). Much of this energy is returned during generation, because power output is strongly related to the head. If water is raised 2 ft (61 cm) by pumping on a high tide of 10 ft (3 m), this will have been raised by 12 ft (3.7 m) at low tide.

Two-Basin Schemes

Another form of energy barrage configuration is that of the dual basin type. With two basins, one is filled at high tide and the other is emptied at low tide. Turbines are placed between the basins. Two-basin schemes offer advantages over normal schemes in that generation time can be adjusted with high flexibility and it is also possible to generate almost continuously. In normal estuarine situations, however, two-basin schemes are very expensive to construct due to the cost of the extra length of barrage. There are some favourable geographies, however, which are well suited to this type of scheme.

Tidal Lagoon Power

Tidal pools are independent enclosing barrages built on high level tidal estuary land that trap the high water and release it to generate power, single pool, around $3.3W/m^2$. Two lagoons operating at different time intervals can guarantee continuous power output, around $4.5W/m^2$. Enhanced pumped storage tidal series of lagoons raises the water level higher than the high tide, and uses intermittent renewables for pumping, around $7.5W/m^2$. i.e. $10 \times 10 km^2$ delivers 750MW constant output 24/7. These independent barrages do not block the flow of the river.

Environmental Impact

The placement of a barrage into an estuary has a considerable effect on the water inside the basin and on the ecosystem. Many governments have been reluctant in recent times to grant approval for tidal barrages. Through research conducted on tidal plants, it has been found that tidal barrages constructed at the mouths of estuaries pose similar environmental threats as large dams. The construction of large tidal plants alters the flow of saltwater in and out of estuaries, which changes the hydrology and salinity and possibly negatively affects the marine mammals that use the estuaries as their habitat The La Rance plant, off the Brittany coast of northern France, was the first and largest tidal barrage plant in the world. It is also the only site where a full-scale evaluation of the ecological impact of a tidal power system, operating for 20 years, has been made

French researchers found that the isolation of the estuary during the construction phases of the tidal barrage was detrimental to flora and fauna, however; after ten years, there has been a "variable degree of biological adjustment to the new environmental conditions"

Some species lost their habitat due to La Rance's construction, but other species colonized the abandoned space, which caused a shift in diversity. Also as a result of the construction, sandbanks disappeared, the beach of St. Servan was badly damaged and high-speed currents have developed near sluices, which are water channels controlled by gates.

Turbidity

Turbidity (the amount of matter in suspension in the water) decreases as a result of smaller volume of water being exchanged between the basin and the sea. This lets light from the Sun penetrate the water further, improving conditions for the phytoplankton. The changes propagate up the food chain, causing a general change in the ecosystem.

Tidal Fences and Turbines

Tidal fences and turbines, if constructed properly, pose less environmental threats than tidal barrages. Tidal fences and turbines, like tidal stream generators, rely entirely on the kinetic motion of the tidal currents and do not use dams or barrages to block channels or estuarine mouths. Unlike barrages, tidal fences do not interrupt fish migration or alter hydrology, thus these options offer energy generating capacity without dire environmental impacts. Tidal fences and turbines can have varying environmental impacts depending on whether or not fences and turbines are constructed with regard to the environment. The main environmental impact of turbines is their impact on fish. If the turbines are moving slowly enough, such as low velocities of 25-50 rpm, fish kill is minimalized and silt and other nutrients are able to flow through the structures For example, a 20 kW tidal turbine prototype built in the St. Lawrence Seaway in 1983 reported no fish kills Tidal fences block off channels, which makes it difficult for fish and wildlife to migrate through those channels. In order to reduce fish kill, fences could be engineered so that the spaces between the caisson wall and the rotor foil are large enough to allow fish to pass through Larger marine mammals such as seals or dolphins can be protected from the turbines by fences or a sonar sensor auto-braking system that automatically shuts the turbines down when marine mammals are detected

Salinity

As a result of less water exchange with the sea, the average salinity inside the basin decreases, also affecting the ecosystem. "Tidal Lagoons" do not suffer from this problem.

Sediment Movements

Estuaries often have high volume of sediments moving through them, from the rivers to the sea. The introduction of a barrage into an estuary may result in sediment accumulation within the barrage, affecting the ecosystem and also the operation of the barrage.

Fish

Fish may move through sluices safely, but when these are closed, fish will seek out turbines and attempt to swim through them. Also, some fish will be unable to escape the water speed near a turbine and will be sucked through. Even with the most fish-friendly turbine design, fish mortality per pass is approximately 15% (from pressure drop, contact with blades, cavitation, etc.). Alternative passage technologies (fish ladders, fish lifts, fish escalators etc.) have so far failed to solve this problem for tidal barrages, either offering extremely expensive solutions, or ones which are used by a small fraction of fish only. Research in sonic guidance of fish is ongoing. The Open-Centre turbine reduces this problem allowing fish to pass through the open centre of the turbine.

Recently a run of the river type turbine has been developed in France. This is a very large slow rotating Kaplan-type turbine mounted on an angle. Testing for fish mortality has indicated fish mortality figures to be less than 5%. This concept also seems very suitable for adaption to marine current/tidal turbines.

Energy Calculations

The energy available from a barrage is dependent on the volume of water. The potential energy contained in a volume of water is:

$$E = \tfrac{1}{2} A \rho g h^2$$

where:

- h is the vertical tidal range,

- A is the horizontal area of the barrage basin,

- ρ is the density of water = 1025 kg per cubic meter (seawater varies between 1021 and 1030 kg per cubic meter) and

- g is the acceleration due to the Earth's gravity = 9.81 meters per second squared.

The factor half is due to the fact, that as the basin flows empty through the turbines, the hydraulic head over the dam reduces. The maximum head is only available at the moment of low water, assuming the high water level is still present in the basin.

Example Calculation of Tidal Power Generation

Assumptions:

- The tidal range of tide at a particular place is 32 feet = 10 m (approx)

- The surface of the tidal energy harnessing plant is 9 km² (3 km × 3 km)= 3000 m × 3000 m = 9 × 10⁶ m²

- Density of sea water = 1025.18 kg/m³

Mass of the sea water = volume of sea water × density of sea water

= (area × tidal range) of water × mass density

= (9 × 10⁶ m² × 10 m) × 1025.18 kg/m³

$= 92 \times 10^9$ kg (approx)

Potential energy content of the water in the basin at high tide = ½ × area × density × gravitational acceleration × tidal range squared

$$= \tfrac{1}{2} \times 9 \times 10^6 \text{ m}^2 \times 1025 \text{ kg/m}^3 \times 9.81 \text{ m/s}^2 \times (10 \text{ m})^2$$

$$= 4.5 \times 10^{12} \text{ J (approx)}$$

Now we have 2 high tides and 2 low tides every day. At low tide the potential energy is zero. Therefore, the total energy potential per day = Energy for a single high tide × 2

$$= 4.5 \times 10^{12} \text{ J} \times 2$$

$$= 9 \times 10^{12} \text{ J}$$

Therefore, the mean power generation potential = Energy generation potential / time in 1 day

$$= 9 \times 10^{12} \text{ J} / 86400 \text{ s}$$

$$= 104 \text{ MW}$$

Assuming the power conversion efficiency to be 30%: The daily-average power generated = 104 MW * 30%

$$= 31 \text{ MW (approx)}$$

Because the available power varies with the square of the tidal range, a barrage is best placed in a location with very high-amplitude tides. Suitable locations are found in Russia, USA, Canada, Australia, Korea, the UK. Amplitudes of up to 17 m (56 ft) occur for example in the Bay of Fundy, where tidal resonance amplifies the tidal range.

Economics

Tidal barrage power schemes have a high capital cost and a very low running cost. As a result, a tidal power scheme may not produce returns for many years, and investors may be reluctant to participate in such projects.

Governments may be able to finance tidal barrage power, but many are unwilling to do so also due to the lag time before investment return and the high irreversible commitment. For example, the energy policy of the United Kingdom recognizes the role of tidal energy and expresses the need for local councils to understand the broader national goals of renewable energy in approving tidal projects. The UK government itself appreciates the technical viability and siting options available, but has failed to provide meaningful incentives to move these goals forward.

Evopod

Evopod is a unique tidal energy device being developed by a UK-based company Oceanflow Energy Ltd for generating electricity from tidal streams and ocean currents. It can operate in exposed deep water sites where severe wind and waves also make up the environment.

1/10 scale Evopod installed in Strangford Lough during 2008.

Floating Tethered Turbines

Advantages

Evopod - Installed in Strangford Lough 2009

- The flow speed in tidal streams and ocean currents tends to be fastest near the surface and falls off in speed as one descend in the water column. As the power that can be extracted from the free flowing water is proportional to the velocity cubed, then a 10% increase in flow speed equates to a 33% increase in power per unit swept area of the turbine.

- The flow is generally more consistent in the top 1/3 of the water column as it is well away from disturbances created by the seabed topography.

- The drag force on turbines of the same power output is proportionally less for a turbine in faster flow (positioned in the upper part of the water column) than a turbine in slower flow (positioned in the lower part of the water column).

- A floating device does not require a flat seabed as the pile anchors require relatively little space and there is no structure on the seabed.

- Turbines supported by floating platforms are more readily accessible for maintenance than those on the seabed.

- Maintaining watertight seals is less problematic for devices positioned higher in the water column as they are not subject to such extreme static pressures.

- Floating devices that are fitted with navigation lights and markings are more readily identifiable under international navigation regulations than unmarked submerged turbines.

Disadvantages

- Floating devices are subject to ocean wave action which can induce motions that will impact on the performance of the turbines they support; semi-submerged devices such as Evopod are designed to be a stable platform in waves so that they can operate for longer and extract more energy from the wave particle velocity. Waves large enough to have an adverse effect on Evopod would affect turbines in all parts of the water column.

- Ocean waves create an orbital movement of the water particles which will add or subtract from the steady ocean current or tidal stream velocity as the wave passes the turbine. Without proper blade pitch or power take-off control systems this could lead to the blade stalling and loss of power output. With proper control systems it is possible to extract this kinetic energy from the waves, much as a wind turbine does in response to wind gusts. The wave particle velocity for short wave reduces with water depth and is therefore less of an issue for deeply submerged turbines. As the wavelength gets longer than it becomes a shallow water wave where there is little change in velocity in the water column.

- The vertical component of the mooring load induced by the drag of the turbine can pull a floating platform under the water unless it is adequately compensated for by change in buoyancy forces, e.g. the submergence of the struts on Evopod, or hydrodynamic lift forces (lifting foils). Tests have shown residual buoyancy in Evopods is sufficient to withstand these forces with the extra bonus of improve system stability.

- Floating devices have to be robust enough to withstand impact from flotsam and in Northern latitudes may need to be designed to cope with ice floes. This is however true for all turbines as flotsam may be fully submerged and therefore impact any seabed turbine.

Design Features

Hull Design and Rotating Midwater Buoy

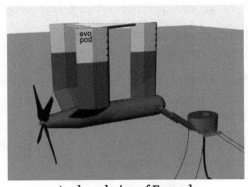

A 3d rendering of Evopod.

The device differentiates itself from other tidal turbines in that the turbine is mounted on a floating, semi-submerged body that is tethered to the seabed. The power generation equipment is similar to that of a wind turbine and is housed in the cylindrical shaped watertight lower hull, which is deeply submerged below the water line and supported by small waterplane area surface piercing struts.

One variant of this patented hull concept has three vertical struts that pierce the water surface, much like a multi-hull SWATH design. The two transversely separated aft struts provide the stability that is needed to resist the torque reacted by the single turbine/generator unit. The configuration of the struts also ensure that the device weathervanes about its midwater mooring buoy such that it always points into the direction of the current.

The device is moored by a mid-water buoy, which is fixed to the seabed by four spread mooring lines which are anchored to the sea-bed by pile or gravity anchors. The buoy design is also unique in that it encompasses a geo-fixed part that is anchored to the seabed and a rotating part that is linked to Evopod by a rigid yoke. The turbine drag forces are therefore transmitted through a bearing system linking the fixed and rotating parts of the buoy. A slip ring power export swivel is located in the buoy so that twist is not imparted into the umbilical cable that takes the power from the midwater buoy to the seabed. A subsea power export cable links the umbilical's seabed connection point to the shore.

With the weather-vaning hull design and rotational midwater buoy, Evopod generates electricity with both the ebb and flood tides by always pointing into the tide's direction of flow. This gives it a generating time of roughly 20 hours per/lunar day (approx 24hrs 50 minutes).

In comparison to other marine bodies that float on the surface of the ocean, Evopod's semi-submerged hull form is hardly affected by the passing waves. It is also designed to be readily detachable from the mid-water buoy for recovery operations. Developing safe installation, maintenance and recovery operations in the hazardous environment of fast flowing currents is one of the biggest challenges facing tidal energy device developers.

The device is designed for deep water sites, such as the Pentland Firth (Up to 60meters water depth, flow speed 6 m/s). Deep water sites in UK waters have the fastest flow speeds and have the greatest potential for electricity generation.

Testing and Collaboration

Evopod 1/40th Scale Tank Testing, Newcastle University, England

A 1/40th scale model of Evopod was initially tested in the test tank of Newcastle University during a proof of concept phase.

1kW Evopod Tidal Test Facility Demonstration, Tees Barrage, England

The 1/10 scale device was initially used to demonstrate the tidal test facilities at the Tees Barrage in Thornaby-on-Tees near Middlesbrough, UK by Narec (National Renewable Energy Centre).

1kW Evopod Sea Testing, Portaferry, Northern Ireland

In 2008 a 1/10 scale Evopod device was installed and tested in the tidal flow through Strangford Narrows near Portaferry, Northern Ireland. Over a period of two years the device collected data but was not connected to the grid under the Supergen Marine Energy Research Programme in collaboration with Queen's University Belfast, amongst others. In 2011 the device was upgraded to include a power export solution which feeds Evopod's generated power onshore to the Queen's University Marine Laboratory. The power is currently fed into the mains circuit of the Marine Laboratory, with plans to be fully grid connected in the near future.

35kW Evopod Sea Testing, Sanda Sound, Scotland

In 2010 Oceanflow Energy were awarded a Scottish WATERS grant to "Build and deploy the 'Evopod', a 35 kilowatt floating grid connected tidal energy turbine at Sanda Sound in South Kintyre".

Awards

Oceanflow Energy and Evopod have won several awards, the most recent being the Shell Springboard Regional award in February 2009. It has also won awards for "innovation of the year" and "green business of the year" in the North East of England.

SeaGen

SeaGen is the world's first large scale commercial tidal stream generator. It was four times more powerful than any other tidal stream generator in the world at the time of installation.

The first SeaGen generator was installed in Strangford Narrows between Strangford and Portaferry in Northern Ireland, UK in April 2008 and was connected to the grid in July 2008. It generates 1.2 MW for between 18 and 20 hours a day while the tides are forced in and out of Strangford Lough through the Narrows. Strangford Lough was also the site of the very first known tide mill in the world, the Nendrum Monastery mill where remains dating from 787 have been excavated.

Background

Marine Current Turbines, the developer of SeaGen, demonstrated first prototype of tidal stream generator in 1994 with a 15 kilowatt system in Loch Linnhe, off the west coast of Scotland. In May 2003, the prototype for SeaGen, 'SeaFlow', was installed off the coast of Lynmouth, North Devon, England. Seaflow was a single rotor turbine which generated 300 kW but was not connected to the grid. SeaFlow was the world's first offshore tidal generator, and remained the world's largest until SeaGen was installed.

The SeaGen rotors can be raised above the surface for maintenance.

SeaGen's predecessor, the 300 kW 'SeaFlow' turbine off the north coast of Devon

Technology

SeaGen generator weighs 300 tonnes. each driving a generator through a gearbox like a hydro-electric or wind turbine. These turbines have a patented feature by which the rotor blades can be pitched through 180 degrees allowing them to operate in both flow directions – on ebb and flood tides. The company claims a capacity factor of 0.59 (average of the last 2000 hours). The power units of each system are mounted on arm-like extensions either side of a tubular steel monopile some 3 metres (9.8 ft) in diameter and the arms with the power units can be raised above the surface for safe and easy maintenance access. The SeaGen was built at Belfast's Harland and Wolff's shipyards.

Environmental Impact

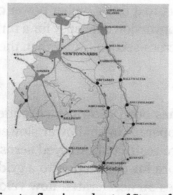

400 million gallons of water flow in and out of Strangford Lough twice a day.

SeaGen has been licensed to operate over a period of 5 years, during which there will be a comprehensive environmental monitoring programme to determine the precise impact on the marine environment.

Problems

During the commissioning of the system a software error caused the blades of one of the turbines to be damaged. This left the turbine operating at half power until autumn 2008. The incident is being investigated and MCT is confident that it will not happen again. Full power operation was finally achieved on 18 Dec 2008.

Shrouded Tidal Turbine

The shrouded turbine Race Rocks Tidal Current Generator before installation at Race Rocks in southern British Columbia in 2006. It operates bi-directionally and has proven to be efficient in contributing to the integrated power system of the area.

The shrouded tidal turbine is an emerging tidal stream technology that has a turbine enclosed in a venturi shaped shroud or duct (ventuduct), producing a sub atmosphere of low pressure behind the turbine. The venturi shrouded turbine is not subject to the Betz limit and allows the turbine to operate at higher efficiencies than the turbine alone by increasing the volume of the flow over the turbine. Claimed improvements vary, from 1.15 to 4 times higher power output than the same turbine minus the shroud. The Betz limit of 59.3% conversion efficiency for a turbine in an open flow still applies, but is applied to the much larger shroud cross-section rather than the small turbine cross-section.

Principles

Considerable commercial interest has been shown in shrouded tidal stream turbines due to the increased power output. They can operate in shallower slower moving water with a smaller turbine at sites where large turbines are restricted. Arrayed across a seaway or in fast flowing rivers, shrouded turbines are cabled to shore for connection to a grid or a community. Alternatively the property of the shroud that produces an accelerated flow velocity across the turbine allows tidal flows formerly too slow for commercial use to be used for energy production.

While the shroud may not be practical in wind, as the next generation of tidal stream turbine design it is gaining more popularity and commercial use. The Tidal Energy Pty Ltd tidal turbine is multidirectional able to face up-stream in any direction and the Lunar Energy turbine bi directional. All tidal stream turbines constantly need to face at the correct angle to the water stream in order to operate. The Tidal Energy Pty Ltd is a unique case with a pivoting base. Lunar Energy use a wide angle diffuser to capture incoming flow that may not be inline with the long axis of the turbine. A shroud can also be built into a tidal fence or barrage increasing the performance of turbines.

Types of Shroud

Not all shrouded turbines are the same - the performance of a shrouded turbine varies with the design of the shroud. Not all shrouded turbines have undergone independent scrutiny of claimed performances, as companies closely guard their respective technologies, so quoted performance figures need to be closely scrutinised. Lunar Energy reports a 15%-25% improvement over the same turbine without the shroud. Shrouded turbines do not operate at maximum efficiency when the shroud does not intercept the current flow at the correct angle, which can occur as currents eddy and swirl, resulting in reduced operational efficiency. At lower turbine efficiencies the extra cost of the shroud must be justified, while at higher efficiencies the extra cost of the shroud has less impact on commercial returns. Similarly the added cost of the supporting structure for the shroud has to be balanced against the performance gained. Yawing (pivoting) the shroud and turbine at the correct angle, so it always faces upstream like a wind sock, can increase turbine performance but may need expensive active devices to turn the shroud into the flow. Passive designs can be incorporated, such as floating the shrouded turbine under a pontoon on a swing mooring, or flying the turbine like a kite under water. One design by Tidal Energy Pty Ltd passively yaws the shrouded turbine using a turntable with a peer reviewed claim of 3.84 (384%) increase in efficiency over the same turbine minus the shroud.

Advantages

- A shroud of suitable geometry can increase the flow velocity across the turbine by 3 to 4 times the open or free stream velocity.

- More power generated means greater returns on investment.

- The number of suitable sites is increased as sites formerly too slow for commercial development become viable.

- Where large cumbersome turbines are not suitable, smaller shrouded turbines can be seabed-mounted in shallow rivers and estuaries allowing safe navigation of the water ways.

- Hidden in a shroud, a turbine is less likely to be damaged by floating debris.

- Bio-fouling is also reduced as the turbine is shaded from natural light in shallow water.

- The increased velocities through the turbine effectively water-blast the shroud throat and turbine clean as organisms are unable to attach at increased velocities.

- Described by one manufacturer as 'eco-benign', tidal stream turbines do not interfere with marine life or the environment and have little or no visual amenity impact.

Disadvantages

- Most shrouded turbines are directional, although one exception is the version off Southern Vancouver Island in British Columbia. One-direction fixed shrouds may not capture flow efficiently - in order for the shroud to produce maximum efficiency to use both flood and ebb tide they need to be yawed like a windmill on a pivot or turntable, or suspended under a pontoon on a marine swing mooring allowing the turbine to always face upstream like a wind sock.

- Shrouded turbine loads are 3 to 4 times those of the open or free stream turbine, so a robust mounting system is necessary. However, this mounting system needs to be designed in such a way as to prevent turbulence being spilled onto the turbine or high-pressure waves occurring near the turbine and detuning performance. Streamlining the mounts and or including structural mounts in the shroud geometry performs two functions, that of supporting the turbine and providing a net benefit of 3 to 4 times the power output.

Tidal Farm

A tidal farm is a group of multiple tidal stream generators assembled in the same location used for production of electric power, similar to that of a wind farm. The low-voltage powerlines from the individual units are then connected to a substation, where the voltage is stepped up with the use of a transformer for distribution through a high voltage transmission system.

Technology

Tidal Farms utilize tidal stream generators that are grouped together to produce electricity. These generators use the moving tides to turn turbines that are very similar the wind turbines used on land. The power of the ocean and the turbines advance technology guarantee a much more predictable energy output then regular wind turbines. The turbines are usually located in areas with high tidal activity in order for the generators to be as efficient as possible. What makes tidal farms unique is that they are set up in groups to allow much more energy production. The generators are connected to substations on shore to transform voltage from high to low, or low to high. These generators can be semi-submersible or fixed into the sea floor, which means they would be out of sight and not an eyesore for the public. The turbines that would be used would be very slow moving due to the density of the water, this is very beneficial to the aquatic life because fish would be able to freely pass through without being in danger of dying. Some turbines can also be used in irrigation canals, rivers, and dam whether the flow of water is fast or slow.

Operators

Scotland is one of the main leaders in the effort to utilize tidal energy as an alternative energy resource. In 2012, Scottish Power installed a 30 ft. turbine off of the Orkney Islands. The currents off of these islands are very fast moving and the tests conducted had shown that the generator produced one megawatt of electricity, enough to power 500 homes. Scotland is also looking to install a more powerful generator off of the Sound of Islay that would be capable of powering in upwards of 5,000 homes once fully operational. In January 2015, production of a 400-megawatt tidal generator was being constructed in Northern Scotland. This generator would be capable of powering 175,000 homes. Ocean power is a clean and efficiency with an energy source that never turns off. Using Tidal farms is a much cleaner and efficient way to produce electricity. One of the drawbacks to tidal farms is marine life and how it will affect it. They would also need to set the tidal farms deep in the ocean where it won't affect fishing boats or large ship passing by. The United States of America has nearly 12,380 miles of coastline, and is undergoing an 18-month study to see how well the tidal farms work *source*. The project could cost up to $10 million, including $2 million on fish monitoring equipment and if everything goes to plan the U.S. could be seeing tidal farms along the US coast. The zero-emission tidal farms could be the way to a cleaner planet, a better future and maybe reducing the cost of electricity in the long run.

Types

- Double and Three Bladed Turbine are turbines attached to a stationary pole and rotate axial to chase the ocean currents. Some of the double and triple bladed turbines can have two sets attached to the pole for better efficiency. This type of turbine has to be detached from the stationary pole and lifter with cranes attached to ships when maintenance is to be performed.

- Semi Submersible Turbines is a more expensive turbine but in the long run it is cheaper and is more cost efficient. The turbines are connecting to a stationary post and the turbine generator can be raised and lowered anytime for maintenance.

- Duct Style Turbine uses duct all the way around the entrance of the turbine to guide and accelerate the tidal stream toward the rotor. By using a duct, more energy can be extracted from the same amount of water with smaller diamter rotor blades thereby keeping costs of manufacture and maintenance down.

- Cable Tethered Turbine floating turbines are attached by a chain to a stationary point in the bottom of the ocean and follow the ocean current in a horizontal 360 degrees. These turbines are easy to bring up to the surface for maintenance because they are pressurized with air. They are also equipped with sensors to detect any water trying to make its way into the pressurized generator.

Problems

One of the few environmental unknowns about tidal farms is the threat they may pose to the plant life in the areas that the turbines would be set up. But by having the blades turn at a slower than normal speed wind turbines can eliminate some of the potential environmental problems. Another problem that can occur is making the turbines water tight to prevent seawater from corrosion the metal parts inside the turbine. Underwater turbines would have to be position away from shipping lanes, too close to shore, and in deep enough waters for them not to interfere with everyday shipping traffic. Having countries like Scotland position underwater turbines can help other countries learn and explore betters ideas for creating energy by learning from the successes and the failures achieved by the leading countries. Scotland expects the rest of the world to follow their example and install tidal farms all over the world in the efforts to help stop pollution and make producing energy cleaner and safer.

Kaipara Tidal Power Station

The Kaipara tidal power station is a proposed tidal power project to be located in the Kaipara Harbour. The project is being developed by Crest Energy, with an ultimate size of 200MW at a cost of $600 million.

Crest plans to place the turbines at least 30 metres deep along a ten kilometre stretch of the main channel. Historical charts show this stretch of the channel has changed little over 150 years. The output of the turbines will cycle twice daily with the predictable rise and fall of the tide. Each turbine will have a maximum output of 1.2 MW, and is expected to generate 0.75 MW averaged over time.

In 2013, it was announced that the project had been put on hold, and most of the shares in Crest Energy had been sold to Todd Energy.

Kaipara Harbour

The entrance to Kaipara Harbour, one of the largest harbours in the world, is a channel to the Tasman Sea. It narrows to a width of 6 kilometres (3.7 mi), and is over 50 metres (160 ft) deep in parts. On average, Kaipara tides rise and fall 2.10 metres (6.9 ft). At high tide, nearly 1000 square kilometres are flooded. Spring tidal flows reach 9 km/h (5 knots) in the entrance channel and move 1,990 million cubic meters per tidal movement or 7,960 million cubic meters daily.

Consents

In 2008, the Northland Regional Council granted resource consents for only 100 turbines. This was appealed to the Environment Court, which in 2011 set conditions for the project allowing for 200 turbines, with many conditions including staged development. The Minister of Conservation granted resource consents for the project in March 2011.

Garorim Bay Tidal Power Station

Garorim Bay

Garorim Bay Tidal Power Station is a planned tidal power plant in Garorim Bay, on the west coast of South Korea. The project is developed by Korea Western Power Company Limited and was in the process of receiving government approval as of November 2008.

Description

Garorim Bay is located between Seosan City and Taean County of Chungnam Province, South Korea, at the western seashore of South Korea. The electric power generation capacity of the plant will be 520 megawatt (26 MW * 20 sets). This is more than twice the capacity of the Rance Power Plant in France.

According to an announcement made by the power company, construction cost was estimated to be 1 trillion Korean won (1 billion US dollars) as of 2005.

History

In 1981, the Assessment Report on the Construction of Tidal Power Plant in Garorim Bay was issued. Environmental research conducted by Korean Ocean Research and Development Institute and supported by UNESCO was partly used in this assessment. The report assessed that the project was economically efficient. But in 1986, as oil prices went down and construction costs rose, the project was assessed as economically inefficient. The economic efficiency was assessed again from 2004 to 2005, and the basic design was completed as of March 2007.

A news reporter reported that the benefit/cost ratio was assessed to be only 0.87 in articles written in 2007.

Environmental Impact

Garorim Bay is regarded as one of the important tidal flats in South Korea. The Korean government included this bay in the National Wetland Inventory. The Yellow Sea Large Marine Ecosystem (YSLME) Project, managed by the United Nations Development Programme and the Global Environment Facility, also surveyed the ecology of this bay.

According to a survey on fishery resources in 1981, supported by the Korean Ocean Research and Development Institute, this bay is an important spawning ground for many species of fish. The study predicted that the plant's construction would cause critical damage to the bay's ecology.

The bay is one of the most important fish farming sites in Korea, composed of about 2000 fishery households.

Response of Local Fishermen

On 20 August 2007, the power company tried to hold a Hearing from Residents, an official proceeding required by the Environmental Impact Assessment (EIA). But about 1000 fishermen, demanding cancellation of the entire project, prevented the hearing from being held so the EIA requirement would not be met.

Relation to Renewable Energy Strategy

Garorim Bay Tidal Power Plant Project is now being included in the South Korean government's renewable energy strategy. The Korean government classifies tidal power as renewable, and the Garorim Bay Project is included in the renewable energy plan which the government announced in September 2008.

The Korean Federation for Environmental Movement (Friends of the Earth Korea) has criticized the project, arguing that the power plant is contrary to the purpose of renewable energy, because it would destroy the valuable tidal flats in the area, thus accelerating global warming.

Rance Tidal Power Station

The Rance Tidal Power Station is a tidal power station located on the estuary of the Rance River in Brittany, France.

Opened in 1966 as the world's first tidal power station, it is currently operated by Électricité de France and was for 45 years the largest tidal power station in the world by installed capacity until the South Korean Sihwa Lake Tidal Power Station surpassed it in 2011.

Its 24 turbines reach peak output at 240 megawatts (MW) and average 57 MW, a capacity factor of approximately 24%. At an annual output of approximately 500 GWh (491 GWh in 2009, 523 GWh in 2010), it supplies 0.12% of the power demand of France. The power density is of the order of 2.6 W/m^2. The cost of electricity production is estimated at €0.12/kWh.

The barrage is 750 m (2,461 ft) long, from Brebis point in the west to Briantais point in the east. The power plant portion of the dam is 332.5 m (1,091 ft) long and the tidal basin measures 22.5 km^2 (9 sq mi).

History

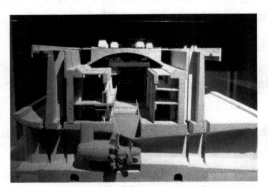
Scale model of the power station

An early attempt to build a tidal power plant was made at Aber Wrac'h in the Finistère in 1925, but due to insufficient finance, it was abandoned in 1930. Plans for this plant served as the draft for follow-on work. Use of tidal energy is not an entirely new concept, since tidal mills have long existed in areas exposed to tides, particularly along the Rance.

The power plant

The idea of constructing a tidal power plant on the Rance dates to Gerard Boisnoer in 1921. The site was attractive because of the wide average-range between low and high tide levels, 8 m (26.2 ft) with a maximum perigean spring tide range of 13.5 m (44.3 ft). The first studies which envisaged a tidal plant on the Rance were done by the Society for the Study of Utilization of the Tides in 1943. Nevertheless, work did not actually commence until 1961. Albert Caquot, the visionary engineer, was instrumental in the construction of the dam, designing an enclosure in order to protect the construction site from the ocean tides and the strong streams. Construction necessitated draining the area where the plant was to be built, which required construction of two dams which took two years. Construction of the plant commenced on 20 July 1963, while the Rance was entirely blocked by the two dams.

Construction took three years and was completed in 1966. Charles de Gaulle, then President of France, inaugurated the plant on 26 November of the same year. Inauguration of the route crossing the plant took place on 1 July 1967, and connection of the plant to the French National Power Grid was carried out on 4 December 1967. In total, the plant cost F620 million (approximately €94.5 million). It took almost 20 years for the La Rance to pay for itself.

Assessments

In spite of the high development cost of the project, the costs have now been recovered, and electricity production costs are lower than that of nuclear power generation (1.8 ¢/kWh versus 2.5 ¢/kWh for nuclear). However, the capacity factor of the plant is 28%, lower than 85-90% for nuclear power.

Environmental Impact

The barrage has caused progressive silting of the Rance ecosystem. Sand-eels and plaice have disappeared, though sea bass and cuttlefish have returned to the river. By definition, tides still flow in the estuary and the operator, EDF, endeavours to adjust their level to minimize the biological impact.

Tourist Attraction

The facility attracted approximately 40,000 visitors in 2011. A canal lock in the west end of the dam permits the passage of 16,000 tonne vessels between the English Channel and the Rance. Departmental highway 168 crosses the dam and allows vehicles to travel between Dinard and

Saint-Malo. There is a drawbridge where the road crosses the lock which may be raised to allow larger vessels to pass.

References

- Turcotte, D. L.; Schubert, G. (2002). "Chapter 4". Geodynamics (2nd ed.). Cambridge, England, UK: Cambridge University Press. pp. 136–137. ISBN 978-0-521-66624-4.

- Elsevier Ltd, The Boulevard, Langford Lane, Kidlington, Oxford, OX5 1GB, United Kingdom. "Green light for world's first tidal lagoon". renewableenergyfocus.com. Retrieved 26 July 2015.

- Lewis, M.; Neill, S.P.; Robins, P.E.; Hashemi, M.R. "Resource assessment for future generations of tidal-stream energy arrays". Energy. 83: 403–415. doi:10.1016/j.energy.2015.02.038.

- Oliver, Antony (9 June 2015). "Swansea Tidal Lagoon power plant wins planning permission". http://www.infrastructure-intelligence.com/. Retrieved 6 July 2015. External link in |work= (help)

- Schweitzer, Sophia. "Will Tidal and Wave Energy Ever Live Up to Their Potential?". Yale Environment 360. Retrieved 16 October 2015.

- Ellie. "First full-scale tidal generator in Wales unveiled: Deltastream array to power 10,000 homes using ebb and flow of the ocean" Daily Mail, 7 August 2014.

- "BBC Look North "A tidal power project in the Humber has generated its first batch of electricity"". Youtube.com. 2009-08-06. Retrieved 2013-04-28.

- Tweed, Katherine (2013-11-14). "Underwater Kite Harvests Energy From Slow Currents - IEEE Spectrum". Spectrum.ieee.org. Retrieved 2013-12-03.

- Supergen Marine Energy Annual Assembly - Prof Trevor Whittaker, Dr Graeme Savage, Dr Matt Folley, Mr Cuan Boake (1-10-2008) "Progress Towards the Sea" (PDF) Retrieved on 21-8-2012

- "1st tidal power delivered to US grid off Maine" Archived September 16, 2012, at the Wayback Machine., CBS MoneyWatch, September 14, 2012

- "Turbines Off NYC East River Will Create Enough Energy to Power 9,500 Homes". U.S. Department of Energy. Retrieved 13 February 2012.

- Carrington, Damian (2011-03-02). "Underwater kite-turbine may turn tides into green electricity | Damian Carrington | Environment". theguardian.com. Retrieved 2013-12-03.

- Yekang Ko and Derek K. Schubert (29 November 2011). "South Korea's Plans for Tidal Power: When a "Green" Solution Creates More Problems" (PDF). report. Nautilus Institute. Retrieved 27 May 2016.

Watermill: A Comprehensive Study

Watermill is the mill that uses hydropower. It uses the water wheel to perform mechanical processes such as rolling and hammering. Poncelet wheel, Sagebien wheel, water turbine, Francis turbine, Kaplan turbine, Pelton wheel and turgo turbine are some of the aspects of watermills that have been elucidated in the following chapter.

Watermill

A watermill or water mill is a mill that uses moving water as its power source. It is thus a structure that uses a water wheel or water turbine to drive a mechanical process such as milling (grinding), rolling, or hammering. Such processes are needed in the production of many material goods, including flour, lumber, paper, textiles, and many metal products. Thus watermills may be gristmills, sawmills, paper mills, textile mills, hammermills, trip hammering mills, rolling mills, wire drawing mills, and so on.

Interior of the Lyme Regis watermill, UK (14th century)

One major way to classify watermills is by wheel orientation (vertical or horizontal), one powered by a vertical waterwheel through a gearing mechanism, and the other equipped with a horizontal waterwheel without such a mechanism. The former type can be further divided, depending on where the water hits the wheel paddles, into undershot, overshot, breastshot and pitchback (backshot or reverse shot) waterwheel mills. Another way to classify water mills is by an essential trait about their location: tide mills use the movement of the tide; ship mills are water mills onboard (and constituting) a ship.

History

Classical Antiquity

Model of a Roman type Roman water-powered grain-mill described by Vitruvius. The millstone (upper floor) is powered by an undershot waterwheel by the way of a gear mechanism (lower floor)

Vertical axle watermill in Dalarna, Sweden.

Hellenistic engineers invented the two main components of watermills, the waterwheel and toothed gearing, and were, along with the Romans, the first to operate undershot, overshot and breastshot waterwheel mills.

The earliest evidence of a water-driven wheel is probably the Perachora wheel (3rd century BC), in Greece. The earliest written reference is in the technical treatises *Pneumatica* and *Parasceuastica* of the Greek engineer Philo of Byzantium (c. 280–220 BC). The British historian of technology M.J.T. Lewis has shown that those portions of Philo of Byzantium's mechanical treatise which describe water wheels and which have been previously regarded as later Arabic interpolations, actually date back to the Greek 3rd-century BC original. The sakia gear is, already fully developed, for the first time attested in a 2nd-century BC Hellenistic wall painting in Ptolemaic Egypt.

Lewis assigns the date of the invention of the horizontal-wheeled mill to the Greek colony of Byzantium in the first half of the 3rd century BC, and that of the vertical-wheeled mill to Ptolemaic Alexandria around 240 BC.

The Greek geographer Strabon reports in his *Geography* a water-powered grain-mill to have existed near the palace of king Mithradates VI Eupator at Cabira, Asia Minor, before 71 BC.

The Roman engineer Vitruvius has the first technical description of a watermill, dated to 40/10 BC; the device is fitted with an undershot wheel and power is transmitted via a gearing mechanism. He also seems to indicate the existence of water-powered kneading machines.

The Greek epigrammatist Antipater of Thessalonica tells of an advanced overshot wheel mill around 20 BC/10 AD. He praised for its use in grinding grain and the reduction of human labour:

Hold back your hand from the mill, you grinding girls; even if the cockcrow heralds the dawn, sleep on. For Demeter has imposed the labours of your hands on the nymphs, who leaping down upon the topmost part of the wheel, rotate its axle; with encircling cogs, it turns the hollow weight of the Nisyrian millstones. If we learn to feast toil-free on the fruits of the earth, we taste again the golden age.

The Roman encyclopedist Pliny mentions in his *Naturalis Historia* of around 70 AD water-powered trip hammers operating in the greater part of Italy. There is evidence of a fulling mill in 73/4 AD in Antioch, Roman Syria.

It is likely that a water-powered stamp mill was used at Dolaucothi to crush gold-bearing quartz, with a possible date of the late 1st century to the early 2nd century. The stamps were operated as a batch of four working against a large conglomerate block, now known as *Carreg Pumpsaint*. Similar anvil stones have been found at other Roman mines across Europe, especially in Spain and Portugal.

The 1st-century AD multiple mill complex of Barbegal in southern France has been described as "the greatest known concentration of mechanical power in the ancient world". It featured 16 overshot waterwheels to power an equal number of flour mills. The capacity of the mills has been estimated at 4.5 tons of flour per day, sufficient to supply enough bread for the 12,500 inhabitants occupying the town of Arelate at that time. A similar mill complex existed on the Janiculum hill, whose supply of flour for Rome's population was judged by emperor Aurelian important enough to be included in the Aurelian walls in the late 3rd century.

A breastshot wheel mill dating to the late 2nd century AD was excavated at Les Martres-de-Veyre, France.

Scheme of the Roman Hierapolis sawmill, the earliest known machine to incorporate a crank and connecting rod mechanism.

The 3rd-century AD Hierapolis water-powered stone sawmill is the earliest known machine to incorporate a crank and connecting rod mechanism. Further sawmills, also powered by crank and connecting rod mechanisms, are archaeologically attested for the 6th-century water-powered stone sawmills at Gerasa and Ephesus. Literary references to water-powered marble saws in what is now Germany can be found in Ausonius 4th-century poem *Mosella*. They also seem to be indicated about the same time by the Christian saint Gregory of Nyssa from Anatolia, demonstrating a diversified use of water-power in many parts of the Roman Empire.

Roman turbine mill at Chemtou, Tunisia. The tangential water inflow of the millrace made the horizontal wheel in the shaft turn like a true turbine, the earliest known.

The earliest turbine mill was found in Chemtou and Testour, Roman North Africa, dating to the late 3rd or early 4th century AD. A possible water-powered furnace has been identified at Marseille, France.

Mills were commonly used for grinding grain into flour (attested by Pliny the Elder), but industrial uses as fulling and sawing marble were also applied.

The Romans used both fixed and floating water wheels and introduced water power to other provinces of the Roman Empire. So-called 'Greek Mills' used water wheels with a horizontal wheel (and vertical shaft). A "Roman Mill" features a vertical wheel (on a horizontal shaft). Greek style mills are the older and simpler of the two designs, but only operate well with high water velocities and with small diameter millstones. Roman style mills are more complicated as they require gears to transmit the power from a shaft with a horizontal axis to one with a vertical axis.

Although to date only a few dozen Roman mills are archaeologically traced, the widespread use of aqueducts in the period suggests that many remain to be discovered. Recent excavations in Roman London, for example, have uncovered what appears to be a tide mill together with a possible sequence of mills worked by an aqueduct running along the side of the River Fleet.

In 537 AD, ship mills were ingeniously used by the East Roman general Belisarius, when the besieging Goths cut off the water supply for those mills. These floating mills had a wheel that was attached to a boat moored in a fast flowing river.

Middle Ages

At the time of the compilation of the Domesday Book (1086), there were 5,624 watermills in England alone, only 2% of which have not been located by modern archeological surveys. Later research estimates a less conservative number of 6,082, and it has been pointed out that this should be considered a minimum as the northern reaches of England were never properly recorded. In 1300, this number had risen to between 10,000 and 15,000. By the early 7th century, watermills were well established in Ireland, and began to spread from the former territory of the empire into the non-romanized parts of Germany a century later. Ship mills and tide mill were introduced in the 6th century.

Tide mills

In recent years, a number of new archaeological finds has consecutively pushed back the date of the earliest tide mills, all of which were discovered on the Irish coast: A 6th-century vertical-wheeled tide mill was located at Killoteran near Waterford. A twin flume horizontal-wheeled tide mill dating to c. 630 was excavated on Little Island. Alongside it, another tide mill was found which was powered by a vertical undershot wheel. The Nendrum Monastery mill from 787 was situated on an island in Strangford Lough in Northern Ireland. Its millstones are 830mm in diameter and the horizontal wheel is estimated to have developed 7–8 hp at its peak. Remains of an earlier mill dated at 619 were also found at the site.

Survey of industrial mills

In a 2005 survey the scholar Adam Lucas identified the following first appearances of various industrial mill types in Western Europe. Noticeable is the preeminent role of France in the introduction of new innovative uses of waterpower. However, he has drawn attention to the dearth of studies of the subject in several other countries.

First Appearance of Various Industrial Mills in Medieval Europe, AD 770-1443		
Type of mill	Date	Country
Malt mill	770	France
Fulling mill	1080	France
Tanning mill	c. 1134	France
Forge mill	c. 1200	England, France
Tool-sharpening mill	1203	France
Hemp mill	1209	France
Paper mill	1238, 1273	Xativa, Spain
Bellows	1269, 1283	Medieval Hungary, France
Sawmill	c. 1300	France
Ore-crushing mill	1317	Germany
Blast furnace	1384	France
Cutting and slitting mill	1443	France

Ancient China

A Northern Song era (960–1127) water-powered mill for dehusking grain with a horizontal wheel.

The waterwheel was found in China from 30 AD onwards, when it was used to power trip hammers, the bellows in smelting iron, and in one case, to mechanically rotate an armillary sphere for astronomical observation. Although Joseph Needham speculates that the water-powered millstone could have existed in Han China by the 1st century AD, there is no sufficient literary evidence for it until the 5th century. In 488 AD, the mathematician and engineer Zu Chongzhi had a watermill erected which was inspected by Emperor Wu of Southern Qi (r. 482–493 AD). The engineer Yang Su of the Sui Dynasty (581–618 AD) was said to operate hundreds of them by the beginning of the 6th century. A source written in 612 AD mentions Buddhist monks arguing over the revenues gained from watermills. The Tang Dynasty (618–907 AD) 'Ordinances of the Department of Waterways' written in 737 AD stated that watermills should not interrupt riverine transport and in some cases were restricted to use in certain seasons of the year. From other Tang-era sources of the 8th century, it is known that these ordinances were taken very seriously, as the government demolished many watermills owned by great families, merchants, and Buddhist abbeys that failed to acknowledge ordinances or meet government regulations. A eunuch serving Emperor Xuanzong of Tang (r. 712–756 AD) owned a watermill by 748 AD which employed five waterwheels that ground 300 bushels of wheat a day. By 610 or 670 AD, the watermill was introduced to Japan via Korean Peninsula. It also became known in Tibet by at least 641 AD.

Ancient India

According to Greek historical tradition, India received water-mills from the Roman Empire in the early 4th century AD when a certain Metrodoros introduced "water-mills and baths, unknown among them [the Brahmans] till then".

Islamic World

Muslim engineers adopted the Greek watermill technology from the Byzantine Empire, where it had been applied for centuries in those provinces conquered by the Muslims, including modern-day Syria, Jordan, Israel, Algeria, Tunisia, Morocco, and Spain.

The industrial uses of watermills in the Islamic world date back to the 7th century, while horizontal-wheeled and vertical-wheeled watermills were both in widespread use by the 9th century. A variety of industrial watermills were used in the Islamic world, including gristmills, hullers, sawmills, shipmills, stamp mills, steel mills, sugar mills, and tide mills. By the 11th century, every province throughout the Islamic world had these industrial watermills in operation, from al-Andalus and North Africa to the Middle East and Central Asia. Muslim and Middle Eastern Christian engineers also used crankshafts and water turbines, gears in watermills and water-raising machines, and dams as a source of water, used to provide additional power to watermills and water-raising machines. Fulling mills, and steel mills may have spread from Al-Andalus to Christian Spain in the 12th century. Industrial watermills were also employed in large factory complexes built in al-Andalus between the 11th and 13th centuries.

An Afghan water mill photographed during the Second Anglo-Afghan War (1878-1880). The rectangular water mill has a thatched roof and traditional design with a small horizontal mill-house built of stone or perhaps mud bricks

The engineers of the Islamic world used several solutions to achieve the maximum output from a watermill. One solution was to mount them to piers of bridges to take advantage of the increased flow. Another solution was the shipmill, a type of watermill powered by water wheels mounted on the sides of ships moored in midstream. This technique was employed along the Tigris and Euphrates rivers in 10th-century Iraq, where large shipmills made of teak and iron could produce 10 tons of flour from corn every day for the granary in Baghdad.

Persia

More than 300 watermills were at work in Iran till 1960. Now only a few are still working. One of the famous ones is the water mill of Askzar and the water mill of the Yazd city, still producing flour.

Operation

A watermill in Tapolca, Veszprem County, Hungary

Roblin's Mill, a watermill, at Black Creek Pioneer Village in Toronto, Canada.

Watermills in Bosnia and Herzegovina.

Typically, water is diverted from a river or impoundment or mill pond to a turbine or water wheel, along a channel or pipe (variously known as a flume, head race, mill race, leat, leet, lade ((Scots) or penstock). The force of the water's movement drives the blades of a wheel or turbine, which in turn rotates an axle that drives the mill's other machinery. Water leaving the wheel or turbine is drained through a tail race, but this channel may also be the head race of yet another wheel, turbine or mill. The passage of water is controlled by sluice gates that allow maintenance and some measure of flood control; large mill complexes may have dozens of sluices controlling complicated interconnected races that feed multiple buildings and industrial processes.

Watermills can be divided into two kinds, one with a horizontal water wheel on a vertical axle, and the other with a vertical wheel on a horizontal axle. The oldest of these were horizontal mills in which the force of the water, striking a simple paddle wheel set horizontally in line with the flow turned a runner stone balanced on the rynd which is atop a shaft leading directly up from the wheel. The bedstone does not turn. The problem with this type of mill arose from the lack of gearing; the speed of the water directly set the maximum speed of the runner stone which, in turn, set the rate of milling.

Most watermills in Britain and the United States of America had a vertical waterwheel, one of four kinds: undershot, breast-shot, overshot and pitchback wheels. This vertical produced rotary motion around a horizontal axis, which could be used (with cams) to lift hammers in a forge, fulling stocks in a fulling mill and so on.

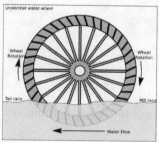

Undershot water wheel, applied for watermilling since the 1st century BC

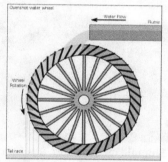

Overshot water wheel, applied for watermilling since the 1st century BC

Milling Corn

However, in corn mills rotation about a vertical axis was required to drive its stones. The horizontal rotation was converted into the vertical rotation by means of gearing, which also enabled the runner stones to turn faster than the waterwheel. The usual arrangement in British and American corn mills has been for the waterwheel to turn a horizontal shaft on which is also mounted a large pit wheel. This meshes with the wallower, mounted on a vertical shaft, which turns the (larger) great spur wheel (mounted on the same shaft). This large face wheel, set with pegs, in turn, turned a smaller wheel (such as a lantern gear) known as a stone nut, which was attached to the shaft that drove the runner stone. The number of runner stones that could be turned depended directly upon the supply of water available. As waterwheel technology improved mills became more efficient, and by the 19th century, it was common for the great spur wheel to drive several stone nuts, so that a single water wheel could drive as many as four stones. Each step in the process increased the gear ratio which increased the maximum speed of the runner stone. Adjusting the sluice gate and thus the flow of the water past the main wheel allowed the miller to compensate for seasonal variations in the water supply. Finer speed adjustment was made during the milling process by *tentering*, that is, adjusting the gap between the stones according to the water flow, the type of grain being milled, and the grade of flour required.

In many mills (including the earliest) the great spur wheel turned only one stone, but there might be several mills under one roof. The earliest illustration of a single waterwheel driving more than one set of stones was drawn by Henry Beighton in 1723 and published in 1744 by J. T. Desaguliers.

Dalgarven Mill, Ayrshire, United Kingdom.

Shipmill on the Mura, Slovenia

Overshot and Pitchback Mills

The overshot wheel was a later innovation in waterwheels and was around two and a half times more efficient than the undershot. The undershot wheel, in which the main water wheel is simply set into the flow of the mill race, suffers from an inherent inefficiency stemming from the fact that the wheel itself, entering the water behind the main thrust of the flow driving the wheel, followed by the lift of the wheel out of the water ahead of the main thrust, actually impedes its own operation. The overshot wheel solves this problem by bringing the water flow to the top of the wheel. The water fills buckets built into the wheel, rather than the simple paddle wheel design of undershot wheels. As the buckets fill, the weight of the water starts to turn the wheel. The water spills out of the bucket on the down side into a spillway leading back to river. Since the wheel itself is set above the spillway, the water never impedes the speed of the wheel. The impulse of the water on the wheel is also harnessed in addition to the weight of the water once in the buckets. Overshot wheels require the construction of a dam on the river above the mill and a more elaborate millpond, sluice gate, mill race and spillway or tailrace.

An inherent problem in the overshot mill is that it reverses the rotation of the wheel. If a miller wishes to convert a breastshot mill to an overshot wheel all the machinery in the mill has to be rebuilt to take account of the change in rotation. An alternative solution was the pitchback or backshot wheel. A launder was placed at the end of the flume on the headrace, this turned the direction of the water without much loss of energy, and the direction of rotation was maintained. Daniels Mill near Bewdley, Worcestershire is an example of a flour mill that originally used a breastshot wheel, but was converted to use a pitchback wheel. Today it operates as a breastshot mill.

A Breastshot waterwheel at Dalgarven Mill, United Kingdom.

Larger water wheels (usually overshot steel wheels) transmit the power from a toothed annular ring that is mounted near the outer edge of the wheel. This drives the machinery using a spur gear mounted on a shaft rather than taking power from the central axle. However, the basic mode of operation remains the same; gravity drives machinery through the motion of flowing water.

Toward the end of the 19th century, the invention of the Pelton wheel encouraged some mill owners to replace over- and undershot wheels with Pelton wheel turbines driven through penstocks.

Tide Mills

A different type of watermill is the tide mill. This mill might be of any kind, undershot, overshot or horizontal but it does not employ a river for its power source. Instead a mole or causeway is built across the mouth of a small bay. At low tide, gates in the mole are opened allowing the bay to fill with the incoming tide. At high tide the gates are closed, trapping the water inside. At a certain point a sluice gate in the mole can be opened allowing the draining water to drive a mill wheel or wheels. This is particularly effective in places where the tidal differential is very great, such as the Bay of Fundy in Canada where the tides can rise fifty feet, or the now derelict village of Tide Mills in the United Kingdom. A working example can be seen at Eling Tide Mill.

Run of the river schemes do not divert water at all and usually involve undershot wheels the mills are mostly on the banks of sizeable rivers or fast flowing streams. Other watermills were set beneath large bridges where the flow of water between the stanchions was faster. At one point London bridge had so many water wheels beneath it that bargemen complained that passage through the bridge was impaired.

Current Status

By the early 20th century, availability of cheap electrical energy made the watermill obsolete in developed countries although some smaller rural mills continued to operate commercially later throughout the century. A few historic mills such as the Newlin Mill and Yates Mill in the USA and The Darley Mill Centre in the UK still operate for demonstration purposes. Small-scale commercial production is carried out in the UK at Daniels Mill, Little Salkeld Mill and Redbournbury Mill.

Watermill in Jahodná (Slovakia).

Some old mills are being upgraded with modern Hydropower technology, for example those worked on by the South Somerset Hydropower Group in the UK.

In some developing countries, watermills are still widely used for processing grain. For example, there are thought to be 25,000 operating in Nepal, and 200,000 in India. Many of these are still of the traditional style, but some have been upgraded by replacing wooden parts with better-designed metal ones to improve the efficiency. For example, the Centre for Rural Technology in Nepal upgraded 2,400 mills between 2003 and 2007.

Applications

Watermill in Caldas Novas, Brazil.

Former watermill in Kohila, Estonia

- Bark mills ground bark, from oak or chestnut trees to produce a coarse powder for use in tanneries.

- Blade mills were used for sharpening newly made blades.

- Blast furnaces, finery forges, and tinplate works were, until the introduction of the steam engine, almost invariably water powered. Furnaces and Forges were sometimes called iron mills.

- Bobbin mills made wooden bobbins for the cotton and other textile industries.

- Carpet mills for making carpets and rugs were sometimes water-powered.

- Cotton mills were driven by water. The power was used to card the raw cotton, and then to drive the spinning mules and ring frames. Steam engines were initially used to increase the water flow to the wheel, then as the industrial revolution progressed, to directly drive the shafts.

- Fulling or *walk* mills were used for a finishing process on woollen cloth.

- Gristmills, or *corn mills*, grind grains into flour.

- Lead was usually smelted in smeltmills prior to the introduction of the cupola (a reverberatory furnace).

- Needle mills for scouring needles during manufacture were mostly water-powered (such as Forge Mill Needle Museum)

- Oil mills for crushing oil seeds might be wind or water-powered

- Paper mills used water not only for motive power, but also required it in large quantities in the manufacturing process.

- Powder mills for making gunpowder - black powder or smokeless powder were usually water-powered.

- Rolling mills shaped metal by passing it between rollers.

- Sawmills cut timber into lumber.

- Slitting mills were used for slitting bars of iron into rods, which were then made into nails.

- Spoke mills turned lumber into spokes for carriage wheels.

- Stamp mills for crushing ore, usually from non-ferrous mines

- Textile mills for spinning yarn or weaving cloth were sometimes water-powered.

Water Wheel

A water wheel is a machine for converting the energy of free-flowing or falling water into useful forms of power, often in a watermill. A water wheel consists of a large wooden or metal wheel, with a number of blades or buckets arranged on the outside rim forming the driving surface. Most commonly, the wheel is mounted vertically on a horizontal axle, but the tub or Norse wheel is mounted horizontally on a vertical shaft. Vertical wheels can transmit power either through the axle or via a ring gear and typically drive belts or gears; horizontal wheels usually directly drive their load.

Water wheels were still in commercial use well into the 20th century, but they are no longer in common use. Prior uses of water wheels include milling flour in gristmills and grinding wood into pulp for papermaking, but other uses include hammering wrought iron, machining, ore crushing and pounding fiber for use in the manufacture of cloth.

An overshot waterwheel standing 42 ft (13 m) high powers the Old Mill at Berry College in Rome, Georgia, USA

Water wheel powering a mine hoist in *De re metallica* (1566)

Some water wheels are fed by water from a mill pond, which is formed when a flowing stream is dammed. A channel for the water flowing to or from a water wheel is called a mill race (also spelled millrace) or simply a "race", and is customarily divided into sections. The race bringing water from the mill pond to the water wheel is a headrace; the one carrying water after it has left the wheel is commonly referred to as a tailrace.

John Smeaton's scientific investigation of the water wheel led to significant increases in efficiency in the mid to late 18th century and supplying much needed power for the Industrial Revolution.

Water wheels began being displaced by the smaller, less expensive and more efficient turbine, developed by Benoît Fourneyron, beginning with his first model in 1827. Turbines are capable of handling high *heads*, or elevations, that exceed the capability of practical-sized waterwheels.

The main difficulty of water wheels is their dependence on flowing water, which limits where they can be located. Modern hydroelectric dams can be viewed as the descendants of the water wheel, as they too take advantage of the movement of water downhill.

History

The two main functions of water wheels were historically water-lifting for irrigation purposes and as a power source. In terms of power source, water wheels can be turned either by human or animal force or by the water current itself. Water wheels come in two basic designs, either equipped with a vertical or a horizontal axle. The latter type can be subdivided, depending on where the water hits the wheel paddles, into overshot, breastshot and undershot wheels.

Greco-Roman World

Engineers of the Hellenistic era Mediterranean region are credited with the development of the water wheel. Mediterranean engineers of the Hellenistic and Roman periods were also the first to use it for both irrigation and as a power source. The technological breakthrough occurred in the technically advanced and scientifically minded Hellenistic period between the 3rd and 1st centuries BCE. This is seen as an evolution of the paddle-driven water-lifting wheels that had appeared in ancient Egypt by the 4th century BCE. According to John Peter Oleson, both the compartmented wheel and the hydraulic Noria appeared in Egypt by the 4th century BCE, with the Sakia being invented there a century later. This is supported by archeological finds at Faiyum, where the oldest archeological evidence of a water-wheel has been found, in the form of a Sakia dating back to the 3rd century BCE. A papyrus dating to the 2nd century BCE also found in Faiyum mentions a water wheel used for irrigation, a 2nd-century BC fresco found at Alexandria depicts a compartmented Sakia, and the writings of Callixenus of Rhodes mention the use of a Sakia in Ptolemaic Egypt during the reign of Ptolemy IV in the late 3rd century BC.

Drainage Wheels

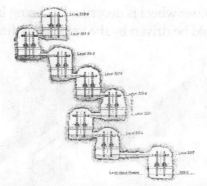

Sequence of wheels found in Rio Tinto mines, southwestern Spain

The Romans used water wheels extensively in mining projects. Several such devices were described by Vitruvius. The one found during modern mining at the copper mines at Rio Tinto in Spain involved 16 such wheels stacked above one another so as to lift water about 80 feet (24 m) from the mine sump. Part of a similar wheel dated to about 90 CE, was found in the 1930s, at Dolaucothi, a Roman gold mine in south Wales.

Water Mills

Reconstruction of Vitruvius' undershot-wheeled watermill

Taking indirect evidence into account from the work of the Greek technician Apollonius of Perge, the British historian of technology M.J.T. Lewis dates the appearance of the vertical-axle watermill to the early 3rd century BCE, and the horizontal-axle watermill to around 240 BC, with Byzantium and Alexandria as the assigned places of invention. A watermill is reported by the Greek geographer Strabon (ca. 64 BCE–CE 24) to have existed sometime before 71 BCE in the palace of the Pontian king Mithradates VI Eupator, but its exact construction cannot be gleaned from the text (XII, 3, 30 C 556).

The first clear description of a geared watermill is from the 1st-century BC Roman architect Vitruvius, who tells of the sakia gearing system as being applied to a watermill. Vitruvius's account is particularly valuable in that it shows how the watermill came about, namely by the combination of the separate Greek inventions of the toothed gear and the water wheel into one effective mechanical system for harnessing water power. Vitruvius's water wheel is described as being immersed with its lower end in the watercourse so that its paddles could be driven by the velocity of the running water (X, 5.2).

Schematic of the Roman Hierapolis sawmill, Asia Minor, powered by a breastshot wheel

About the same time, the overshot wheel appears for the first time in a poem by Antipater of Thessalonica, which praises it as a labour-saving device (IX, 418.4–6). The motif is also taken up by

Lucretius (ca. 99-55 BC) who likens the rotation of the water wheel to the motion of the stars on the firmament (V 516). The third horizontal-axled type, the breastshot water wheel, comes into archaeological evidence by the late-2nd-century AD context in central Gaul. Most excavated Roman watermills were equipped with one of these wheels which, although more complex to construct, were much more efficient than the vertical-axle water wheel. In the 2nd century AD, Barbegal watermill complex a series of sixteen overshot wheels was fed by an artificial aqueduct, a proto-industrial grain factory which has been referred to as "the greatest known concentration of mechanical power in the ancient world".

In Roman North Africa, several installations from around 300 AD were found where vertical-axle water wheels fitted with angled blades were installed at the bottom of a water-filled, circular shaft. The water from the mill-race which entered the pit tangentially created a swirling water column that made the fully submerged wheel act like true water turbines, the earliest known to date.

Navigation

Ox-powered Roman paddle wheel boat from a 15th-century copy of *De Rebus Bellicis*

Apart from its use in milling and water-raising, ancient engineers applied the paddled water wheel for automatons and in navigation. Vitruvius (X 9.5-7) describes multi-geared paddle wheels working as a ship odometer, the earliest of its kind. The first mention of paddle wheels as a means of propulsion comes from the 4th–5th-century military treatise *De Rebus Bellicis* (chapter XVII), where the anonymous Roman author describes an ox-driven paddle-wheel warship.

Early Medieval Europe

Ancient water-wheel technology continued unabated in the early medieval period where the appearance of new documentary genres such as legal codes, monastic charters, but also hagiography was accompanied with a sharp increase in references to watermills and wheels.

The earliest vertical-wheel in a tide mill is from 6th-century Killoteran near Waterford, Ireland, while the first known horizontal-wheel in such a type of mill is from the Irish Little Island (c. 630). As for the use in a common Norse or Greek mill, the oldest known horizontal-wheels were excavated in the Irish Ballykilleen, dating to c. 636.

The earliest excavated water wheel driven by tidal power was the Nendrum Monastery mill in Northern Ireland which has been dated at 787A.D. although a possible earlier mill dates to 619A.D. Tide mills became common in estuaries with a good tidal range in both Europe and America generally using undershot wheels.

Water wheel powering a small village mill at the Museum of Folk Architecture and Life, Uzhhorod, Ukraine

Cistercian monasteries, in particular, made extensive use of water wheels to power watermills of many kinds. An early example of a very large water wheel is the still extant wheel at the early 13th century Real Monasterio de Nuestra Senora de Rueda, a Cistercian monastery in the Aragon region of Spain. Grist mills (for corn) were undoubtedly the most common, but there were also sawmills, fulling mills and mills to fulfil many other labour-intensive tasks. The water wheel remained competitive with the steam engine well into the Industrial Revolution. At around the 8th to 10th century, a number of irrigation technologies were brought into Spain and thus introduced to Europe. One of those technologies is the Noria, which is basically a wheel fitted with buckets on the peripherals for lifting water. It is similar to the undershot water wheel mentioned later in this article. It allowed peasants to power watermills more efficiently. According to Thomas Glick's book, *Irrigation and Society in Medieval Valencia*, the Noria probably originated from somewhere in Persia. It has been used for centuries before the technology was brought into Spain by Arabs who had adopted it from the Romans. Thus the distribution of the Noria in the Iberian peninsula "conforms to the area of stabilized Islamic settlement". This technology has a profound effect on the life of peasants. The Noria is relatively cheap to build. Thus it allowed peasants to cultivate land more efficiently in Europe. Together with the Spaniards, the technology then spread to North Africa and later to the New World in Mexico and South America following Spanish expansion.

Domesday Inventory of English Mills Ca. 1086

The assembly convened by William of Normandy, commonly referred to as the "Domesday" or Doomsday survey, took an inventory of all potentially taxable property in England, which included over six thousand mills spread across three thousand different locations.

Locations

The type of water wheel selected was dependent upon the location. Generally if only small volumes of water and high waterfalls were available a millwright would choose to use an overshot wheel. The decision was influenced by the fact that the buckets could catch and use even a small volume of water. For large volumes of water with small waterfalls the undershot wheel would have been used, since it was more adapted to such conditions and cheaper to construct. So long as these water supplies were abundant the question of efficiency remained irrelevant. By the 18th century with increased demand for power coupled with limited water locales, an emphasis was made on efficiency scheme.

Economic Influence

By the 11th century there were parts of Europe where the exploitation of water was commonplace. The water wheel is understood to have actively shaped and forever changed the outlook of Westerners. Europe began to transit from human and animal muscle labor towards mechanical labor with the advent of the water wheel. Medievalist Lynn White Jr. contended that the spread of inanimate power sources was eloquent testimony to the emergence of the West of a new attitude toward, power, work, nature, and above all else technology.

Harnessing water-power enabled gains in agricultural productivity, food surpluses and the large scale urbanization starting in the 11th century. The usefulness of water power motivated European experiments with other power sources, such as wind and tidal mills. Waterwheels influenced the construction of cities, more specifically canals. The techniques that developed during this early period such as stream jamming and the building of canals, put Europe on a hydraulically focused path, for instance water supply and irrigation technology was combined to modify supply power of the wheel. Illustrating the extent to which there was a great degree of technological innovation that met the growing needs of the feudal state.

Applications of the Water Wheel in Medieval Europe

Ore stamp mill (behind worker taking ore form chute). From Georg Agricola's *De re metallica* (1556)

The water mill was used for grinding grain, producing flour for bread, malt for beer, or coarse meal for porridge. Hammermills used the wheel to operate hammers. One type was fulling mill, which was used for cloth making. The trip hammer was also used for making wrought iron and for working iron into useful shapes, an activity that was otherwise labour-intensive. The water wheel was also used in papermaking, beating material to a pulp. In the 13th century water mills used for hammering throughout Europe improved the productivity of early steel manufacturing. Along with the mastery of gunpowder, waterpower provided European countries worldwide military leadership from the 15th century.

Importance to 17th- and 18th-century Europe (Scientific Influence)

Millwrights distinguished between the two forces, impulse and weight, at work in water wheels long before 18th-century Europe. Fitzherbert, a 16th-century agricultural writer, wrote "druieth the wheel as well as with the weight of the water as with strengthe [impulse]." Leonardo da Vinci also discussed water power, noting "the blow [of the water] is not weight, but excites a power of weight, almost equal to its own power." However, even realisation of the two forces, weight and impulse, confusion remained over the advantages and disadvantages of the two, and there was no clear understanding of the superior efficiency of weight. Prior to 1750 it was unsure as to which force was dominant and was widely understood that both forces were operating with equal inspiration amongst one another. The waterwheel, sparked questions of the laws of nature, specifically the laws of force. Evangelista Torricelli's work on water wheels used an analysis of Galileo's work on falling bodies, that the velocity of a water sprouting from an orifice under its head was exactly equivalent to the velocity a drop of water acquired in falling freely from the same height.

Industrial European Usage

Lady Isabella Wheel, Laxey, Isle of Man, used to drive mine pumps

A mid-19th-century water wheel at Cromford in England used for grinding locally mined barytes.

The most powerful water wheel built in the United Kingdom was the 100 hp Quarry Bank Mill water wheel near Manchester. A high breastshot design, it was retired in 1904 and replaced with several turbines. It has now been restored and is a museum open to the public.

The biggest working water wheel in mainland Britain has a diameter of 15.4 m and was built by the De Winton company of Caernarfon. It is located within the Dinorwic workshops of the National Slate Museum in Llanberis, North Wales.

The largest working water wheel in the world is the Laxey Wheel (also known as *Lady Isabella*) in the village of Laxey, Isle of Man. It is 72 feet 6 inches (22.10 m) in diameter and 6 feet (1.83 m) wide and is maintained by Manx National Heritage.

Development of water turbines during the Industrial Revolution led to decreased popularity of water wheels. The main advantage of turbines is that its ability to harness head is much greater than the diameter of the turbine, whereas a water wheel cannot effectively harness head greater than its diameter. The migration from water wheels to modern turbines took about one hundred years.

China

Two types of hydraulic-powered chain pumps from the *Tiangong Kaiwu* of 1637, written by the Ming Dynasty encyclopedist, Song Yingxing (1587–1666).

Chinese water wheels almost certainly have a separate origin, as early ones there were invariably horizontal water wheels. By at least the 1st century AD, the Chinese of the Eastern Han Dynasty were using water wheels to crush grain in mills and to power the piston-bellows in forging iron ore into cast iron.

In the text known as the *Xin Lun* written by Huan Tan about 20 AD (during the usurpation of Wang Mang), it states that the legendary mythological king known as Fu Xi was the one responsible for the pestle and mortar, which evolved into the tilt-hammer and then trip hammer device. Although the author speaks of the mythological Fu Xi, a passage of his writing gives hint that the water wheel was in widespread use by the 1st century AD in China (Wade-Giles spelling):

Fu Hsi invented the pestle and mortar, which is so useful, and later on it was cleverly improved in such a way that the whole weight of the body could be used for treading on the tilt-hammer (*tui*), thus increasing the efficiency ten times. Afterwards the power of animals—donkeys, mules, oxen, and horses—was applied by means of machinery, and water-power too used for pounding, so that the benefit was increased a hundredfold.

In the year 31 AD, the engineer and Prefect of Nanyang, Du Shi (d. 38), applied a complex use of the water wheel and machinery to power the bellows of the blast furnace to create cast iron. Du Shi is mentioned briefly in the *Book of Later Han* (*Hou Han Shu*) as follows (in Wade-Giles spelling):

In the seventh year of the Chien-Wu reign period (31 AD) Tu Shih was posted to be Prefect of Nanyang. He was a generous man and his policies were peaceful; he destroyed evil-doers and established the dignity (of his office). Good at planning, he loved the common people and wished to save their labor. He invented a water-power reciprocator (*shui phai*) for the casting of (iron) agricultural implements. Those who smelted and cast already had the push-bellows to blow up their charcoal fires, and now they were instructed to use the rushing of the water (*chi shui*) to operate it ... Thus the people got great benefit for little labor. They found the 'water(-powered) bellows' convenient and adopted it widely.

Water wheels in China found practical uses such as this, as well as extraordinary use. The Chinese inventor Zhang Heng (78–139) was the first in history to apply motive power in rotating the astronomical instrument of an armillary sphere, by use of a water wheel. The mechanical engineer Ma Jun (c. 200–265) from Cao Wei once used a water wheel to power and operate a large mechanical puppet theater for the Emperor Ming of Wei (r. 226-239).

India

The early history of the watermill in India is obscure. Ancient Indian texts dating back to the 4th century BC refer to the term *cakkavattaka* (turning wheel), which commentaries explain as *arahatta-ghati-yanta* (machine with wheel-pots attached). On this basis, Joseph Needham suggested that the machine was a noria. Terry S. Reynolds, however, argues that the "term used in Indian texts is ambiguous and does not clearly indicate a water-powered device." Thorkild Schiøler argued that it is "more likely that these passages refer to some type of tread- or hand-operated water-lifting device, instead of a water-powered water-lifting wheel."

According to Greek historical tradition, India received water-mills from the Roman Empire in the early 4th century AD when a certain Metrodoros introduced "water-mills and baths, unknown among them [the Brahmans] till then". Irrigation water for crops was provided by using water raising wheels, some driven by the force of the current in the river from which the water was being raised. This kind of water raising device was used in ancient India, predating, according to Pacey, its use in the later Roman Empire or China, even though the first literary, archaeological and pictorial evidence of the water wheel appeared in the Hellenistic world.

Around 1150, the astronomer Bhaskara Achārya observed water-raising wheels and imagined such a wheel lifting enough water to replenish the stream driving it, effectively, a perpetual motion machine. The construction of water works and aspects of water technology in India is described in Arabic and Persian works. During medieval times, the diffusion of Indian and Persian irrigation technologies gave rise to an advanced irrigation system which brought about economic growth and also helped in the growth of material culture.

Islamic World

Arab engineers took over the water technology of the hydraulic societies of the ancient Near East; they adopted the Greek water wheel as early as the 7th century, excavation of a canal in the Basra region discovered remains of a water wheel dating from this period. Hama in Syria still preserves some of its large wheels, on the river Orontes, although they are no longer in use. One of the largest had a diameter of about 20 metres and its rim was divided into 120 compartments. Another wheel that is still in operation is found at Murcia in Spain, La Nora, and although the original wheel has been replaced by a steel one, the Moorish system during al-Andalus is otherwise virtually unchanged. Some medieval Islamic compartmented water wheels could lift water as high as 30 meters. Muhammad ibn Zakariya al-Razi's *Kitab al-Hawi* in the 10th century described a noria in Iraq that could lift as much as 153,000 litres per hour, or 2550 litres per minute. This is comparable to the output of modern norias in East Asia, which can lift up to 288,000 litres per hour, or 4800 litres per minute.

The norias of Hama on the Orontes River

Water wheel in Djambi, Sumatra, c. 1918

The industrial uses of watermills in the Islamic world date back to the 7th century, while horizontal-wheeled and vertical-wheeled water mills were both in widespread use by the 9th century. A variety of industrial watermills were used in the Islamic world, including gristmills, hullers, sawmills, shipmills, stamp mills, steel mills, sugar mills, and tide mills. By the 11th century, every province throughout the Islamic world had these industrial watermills in operation, from al-Andalus and North Africa to the Middle East and Central Asia. Muslim and Christian engineers also used crankshafts and water turbines, gears in watermills and water-raising machines, and dams as a source of water, used to provide additional power to watermills and water-raising machines. Fulling mills and steel mills may have spread from Islamic Spain to Christian Spain in the 12th century. Industrial water mills were also employed in large factory complexes built in al-Andalus between the 11th and 13th centuries.

The engineers of the Islamic world developed several solutions to achieve the maximum output from a water wheel. One solution was to mount them to piers of bridges to take advantage of the increased flow. Another solution was the shipmill, a type of water mill powered by water wheels mounted on the sides of ships moored in midstream. This technique was employed along the Tigris and Euphrates rivers in 10th-century Iraq, where large shipmills made of teak and iron could produce 10 tons of flour from corn every day for the granary in Baghdad. The flywheel mechanism, which is used to smooth out the delivery of power from a driving device to a driven machine, was invented by Ibn Bassal (fl. 1038-1075) of Al-Andalus; he pioneered the use of the flywheel in the saqiya (chain pump) and noria. The engineers Al-Jazari in the 13th century and Taqi al-Din in the 16th century described many inventive water-raising machines in their technological treatises. They also employed water wheels to power a variety of devices, including various water clocks and automata.

Types

Most water wheels in the United Kingdom and the United States are (or were) vertical wheels rotating about a horizontal axle, but in the Scottish highlands and parts of southern Europe mills often had a horizontal wheel (with a vertical axle). Water wheels are classified by the way in which water is applied to the wheel, relative to the wheel's axle. Overshot and pitchback water wheels are suitable where there is a small stream with a height difference of more than 2 meters, often in association with a small reservoir. Breastshot and undershot wheels can be used on rivers or high volume flows with large reservoirs.

Horizontal Wheel

Commonly called a tub wheel or Norse mill, the horizontal wheel is essentially a very primitive and inefficient form of the modern turbine. It is usually mounted inside a mill building below the working floor. A jet of water is directed on to the paddles of the water wheel, causing them to turn; water exits beneath the wheel, generally through the center. This is a simple system, usually used without gearing so that the vertical axle of the water wheel becomes the drive spindle of the mill.

Undershot Wheel

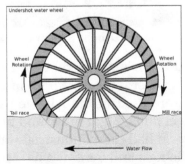

Undershot water wheel

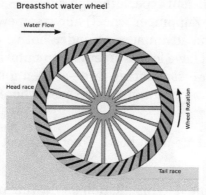

Breastshot water wheel

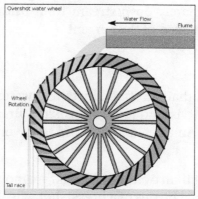

Overshot water wheel

An undershot wheel (also called a *stream wheel*) is a vertically mounted water wheel that is rotated by water striking paddles or blades at the bottom of the wheel. The name *undershot* comes from this striking at the bottom of the wheel. This type of water wheel is the oldest type of wheel.

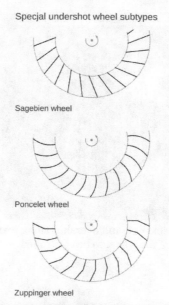

Sabegien, Poncelet and Zuppinger water wheel

It is also regarded as the least efficient type, although subtypes of this water wheel (e.g. the Poncelet wheel, Sagebien wheel and Zuppinger wheel) allow somewhat greater efficiencies than the traditional undershot wheels. The advantages of undershot wheels are that they are somewhat cheaper and simpler to build, and have less of an environmental impact—as they do not constitute a major change of the river. Their disadvantages are—as mentioned before—less efficiency, which means that they generate less power and can only be used where the flow rate is sufficient to provide torque.

Undershot wheels gain no advantage from head. They are most suited to shallow streams in flat country.

Undershot wheels are also well suited to installation on floating platforms. The earliest were probably constructed by the Byzantine general Belisarius during the siege of Rome in 537. Later they were sometimes mounted immediately downstream from bridges where the flow restriction of arched bridge piers increased the speed of the current.

Breastshot Wheel

A vertically mounted water wheel that is rotated by falling water striking buckets near the center of the wheel's edge, or just above it, is said to be *breastshot*. Breastshot wheels are the most common type in the United States of America and are said to have powered the American industrial revolution.

Breastshot wheels are less efficient than overshot wheels, are more efficient than undershot wheels, and are not backshot. The individual blades of a breastshot wheel are actually buckets, as are those of most overshot wheels, and not simple paddles like those of most undershot wheels. A breastshot wheel requires a good trash rack and typically has a masonry "apron" closely conforming to the wheel face, which helps contain the water in the buckets as they progress downwards. Breastshot wheels are preferred for steady, high-volume flows such as are found on the fall line of the North American East Coast.

The Anderson Mill of Texas is undershot, backshot, and overshot using two sources of water. This allows the speed of the wheel to be controlled.

Overshot Wheel

A vertically mounted water wheel that is rotated by falling water striking paddles, blades or buckets near the top of the wheel is said to be *overshot*. In true overshot wheels the water passes over the top of the wheel, but the term is sometimes applied to backshot or pitchback wheels where the water goes down behind the water wheel.

A typical overshot wheel has the water channeled to the wheel at the top and slightly beyond the axle. The water collects in the buckets on that side of the wheel, making it heavier than the other "empty" side. The weight turns the wheel, and the water flows out into the tail-water when the wheel rotates enough to invert the buckets. The overshot design can use all of the water flow for power (unless there is a leak) and does not require rapid flow.

Unlike undershot wheels, overshot wheels gain a double advantage from gravity. Not only is the momentum of the flowing water partially transferred to the wheel, the weight of the water descending in the wheel's buckets also imparts additional energy. The mechanical power derived from an overshot wheel is determined by the wheel's physical size and the available head, so they are ideally suited to hilly or mountainous country. On average, the undershot wheel uses 22 percent of the energy in the flow of water, while an overshot wheel uses 63 percent, as calculated by English civil engineer John Smeaton in the 18th century.

Overshot wheels demand exact engineering and significant head, which usually means significant investment in constructing a dam, millpond and waterways. Sometimes the final approach of the water to the wheel is along a lengthy flume or penstock.

Reversible Wheel

Replica of a reversible wheel with a 9.5 m diameter in Clausthal-Zellerfeld

A special type of overshot wheel is the reversible water wheel. This has two sets of blades or buckets running in opposite directions, so that it can turn in either direction depending on which side the water is directed. Reversible wheels were used in mining industry in order to power various means of ore conveyance. By changing the direction of the wheel, barrels or baskets of ore could be lifted up or lowered down a shaft. As a rule there was also a cable drum or a chain basket (German: *Kettenkorb*) on the axle of the wheel. It was also essential that the wheel had braking equipment in order to be able to stop the wheel (known as a braking wheel). The oldest known drawing of a reversible water wheel was by Georgius Agricola and dates to 1556.

Backshot Wheel

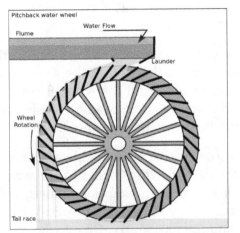

Pitchback or "backshot" water wheel

Backshot wheel at New Lanark World Heritage Site, Scotland

A backshot wheel (also called *pitchback*) is a variety of overshot wheel where the water is introduced just behind the summit of the wheel. It combines the advantages from breastshot and overshot systems, since the full amount of the potential energy released by the falling water is harnessed as the water descends the back of the wheel (as in overshot wheel) while it also gains power from the water's current past the bottom of the wheel (as in breastshot wheel).

A backshot wheel continues to function until the water in the wheel pit rises well above the height of the axle, when any other overshot wheel will be stopped or even destroyed. This makes the technique particularly suitable for streams that experience extreme seasonal variations in flow, and reduces the need for complex sluice and tail race configurations.

The direction of rotation of a backshot wheel is the same as that of a breastshot wheel at the same location so it can easily replace one, without causing the directional gearing in the mill to be changed. This would increase the power available while only requiring a change to be made to the water level in the top pond, which in some cases is economically viable.

Suspension Wheels and Rim-gears

Two early improvements were suspension wheels and rim gearing. Suspension wheels are constructed in the same manner as a bicycle wheel, the rim being supported under tension from the hub- this led to larger lighter wheels than the former design where the heavy spokes were under compression. Rim-gearing entailed adding a notched wheel to the rim or shroud of the wheel. A stub gear engaged the rim-gear and took the power into the mill using an independent line shaft. This removed the rotative stress from the axle which could thus be lighter, and also allowed more flexibility in the location of the power train. The shaft rotation was geared up from that of the wheel which led to less power loss. An example of this design pioneered by Thomas Hewes and refined by William Fairburn can be seen at the 1849 restored wheel at the Portland Basin Canal Warehouse.

The suspension wheel with rim-gearing at the Portland Basin Canal Warehouse

Efficiency

Overshot (and particularly backshot) wheels are the most efficient type; a backshot steel wheel can be more efficient (about 60%) than all but the most advanced and well-constructed turbines. In some situations an overshot wheel is preferable to a turbine.

The development of the hydraulic turbine wheels with their improved efficiency (>67%) opened up an alternative path for the installation of water wheels in existing mills, or redevelopment of abandoned mills.

Power Calculations

The great water wheel in the Welsh National Slate Museum

In an undershot wheel or a run of the river wheel the power is dependant to the kinetic energy of the river. Approximate power can be calculated.

Power in Watts= $100 \times A \times V^3 \times C$

A = Area of paddles in the water (square meters)

V = Velocity of the stream in meters per second

C = Efficiency Constant (assume 1 for a water to wire efficiency of 20%)

Rotational speed of the wheel = 9 × V /D rpm

D = diameter in meters

For a breast shot or over shot wheel both potential energy and kinetic energy must be considered. This takes the form of the weight of water in the buckets and the vertical distance travelled. A rule of thumb formula is

Power in Watts = 4 × Q × H × C

Q = Weight of water (volume per sec x capacity of the buckets)

V = Velocity of the stream in meters per second

H = Head, or height difference of water between the lip of the flume (head race) and the tailrace

C = Efficiency Constant

The optimal rotational speed of a breast shot or overshot wheel is approximately:

Rotational speed of the wheel= 21/ √D

D = diameter of the wheel in metres

Hydraulic Wheel

A recent development of the breastshot wheel is a hydraulic wheel which effectively incorporates automatic regulation systems. The Aqualienne is one example. It generates between 37 kW and 200 kW of electricity from a 20m³ waterflow with a head of 1 to 3.5m. It is designed to produce electricity at the sites of former watermills.

Hydraulic Wheel Part Reaction Turbine

A parallel development is the hydraulic wheel/part reaction turbine that also incorporates a weir into the centre of the wheel but uses blades angled to the water flow. The WICON-Stem Pressure Machine (SPM) exploits this flow. Estimated efficiency 67%.

The University of Southampton School of Civil Engineering and the Environment in the UK has investigated both types of Hydraulic wheel machines and has estimated their hydraulic efficiency and suggested improvements, i.e. The Rotary Hydraulic Pressure Machine. (Estimated maximum efficiency 85%).

These type of water wheels have high efficiency at part loads / variable flows and can operate at very low heads, < 1 metre. Combined with direct drive Axial Flux Permanent Magnet Alternators and power electronics they offer a viable alternative for low head hydroelectric power generation.

Water-lifting

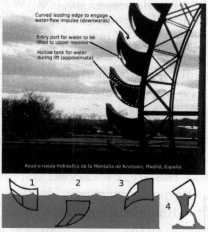

Curved leading edge to engage
water-flow impulse (downwards)

Entry port for water to be
lifted to upper resovoir

Hollow tank for water
during lift (approximate)

Azud o rueda hidráulica de la Montaña de Arunjuez, Madrid, España

Detail of *azud* at Aranjuez, Spain

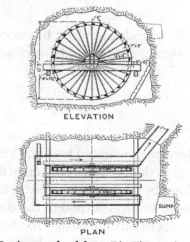

ELEVATION

PLAN

Drainage wheel from Rio Tinto mines

In water-raising devices rotary motion is typically more efficient than machines based on oscillating motion.

The compartmented water wheel comes in two basic forms, the wheel with compartmented body (Latin *tympanum*) and the wheel with compartmented rim or a rim with separate, attached containers. The wheels could be either turned by the flow of water, men treading on its outside or by animals by means of a sakia gear. While the tympanum had a large discharge capacity, it could lift the water only to less than the height of its own radius and required a large torque for rotating. These constructional deficiencies were overcome by the wheel with a compartmented rim which was a less heavy design with a higher lift.

Ptolemaic Egypt

The earliest literary reference to a water-driven, compartmented wheel appears in the technical treatise *Pneumatica* (chap. 61) of the Greek engineer Philo of Byzantium (ca. 280–220 BC). In his *Parasceuastica* (91.43–44), Philo advises the use of such wheels for submerging siege mines as a defensive measure against enemy sapping. Compartmented wheels appear to have been the means of choice for draining dry docks in Alexandria under the reign of Ptolemy IV (221–205 BC). Several Greek papyri of the 3rd to 2nd century BC mention the use of these wheels, but don't give further details. The non-existence of the device in the Ancient Near East before Alexander's conquest can be deduced from its pronounced absence from the otherwise rich oriental iconography on irrigation practices. Unlike other water-lifting devices and pumps of the period though, the invention of the compartmented wheel cannot be traced to any particular Hellenistic engineer and may have been made in the late 4th century BC in a rural context away from the metropolis of Alexandria.

The earliest depiction of a compartmented wheel is from a tomb painting in Ptolemaic Egypt which dates to the 2nd century BC. It shows a pair of yoked oxen driving the wheel via a sakia gear, which is here for the first time attested, too. The Greek sakia gear system is already shown fully developed to the point that "modern Egyptian devices are virtually identical". It is assumed that the scientists of the Museum of Alexandria, at the time the most active Greek research center, may have been involved in its invention. An episode from the Alexandrian War in 48 BC tells of how Caesar's enemies employed geared water wheels to pour sea water from elevated places on the position of the trapped Romans.

Around 300 AD, the noria was finally introduced when the wooden compartments were replaced with inexpensive ceramic pots that were tied to the outside of an open-framed wheel.

Poncelet Wheel

The Poncelet wheel is a type of waterwheel invented by Jean-Victor Poncelet while working at the École d'Application in Metz. It roughly doubled the efficiency of existing undershot waterwheels through a series of detail improvements. The first Poncelet wheel was constructed in 1838, and the design quickly became common in France. Although the design was a great improvement on existing designs, further improvements in turbine design rendered the Poncelet wheel obsolete by the mid-century.

Design

Traditional undershot waterwheels consist of a series of flat blades fixed to the rim of a wheel. The blades were typically mounted so they faced straight out along the radius of the wheel. When water from the millstock flowed past the wheel, it hit the blades, and some of its momentum was transferred to the wheel. However, much of it was also reflected off the blade and lost as heat. This process was not efficient; much of the original velocity in the water remained in it, meaning that potential energy was not being captured. Typical undershot wheels were around 30% efficient.

Jean Charles de Borda was the first to directly characterize the efficiency of waterwheels by comparing the velocities of water before and after meeting the wheel. Poncelet was familiar with this work and started looking for ways to improve the design. He stated that "After having reflected on this, it seemed to me that we could fulfil this double condition by replacing the straight blades on ordinary wheels with curved or cylindrical blades, presenting their concavity to the current."

His design used curved blades positioned so the water met the blade flat to its edge instead of the side. This eliminated the "bounce" that robbed power from the typical design. The water rose up into the channel between the blades for about 15 degrees of rotation, and then drained back out after another 15 degrees, where it dropped out of the channel, over the curve of the blade, imparting further impulse. By the time it left, the water had almost no velocity left. He estimated that practical wheels would reach as high as 80% for low velocity streams, and 70% for high velocity ones that fill the buckets too quickly.

Poncelet developed the design in 1823 and built a small model in 1824 that demonstrated 72% efficiency. Several commercial models followed, including a large installation in Metz that delivered 33% more power than the traditional wheel it replaced, in spite of implementing only some of the design. He published a longer paper on the design in 1826, and a much more detailed version in 1827. The design won a Prix de Mecanique from the French Academy of Sciences, who were funding development of the waterwheel and also awarded several other designs similar awards.

Poncelet wheels became common in France and Germany, where undershot designs were common. However, the large-scale installation of steam engines and water turbines led to the Poncelet wheel falling from use.

Sagebien Wheel

The Sagebien wheel is a type of water wheel invented by Alphonse Sagebien of France, a hydrological engineer and a graduate of the École Centrale des Arts et Manufactures. It was one of the most efficient breastshot water wheel designs of its era; when working on a low head of water, the Sagebien wheel could reach efficiencies of up to 90% in real-world examples.

Design

Traditional breastshot waterwheels consist of a series of flat blades fixed to the rim of a wheel. The blades were typically mounted so they faced straight out along the radius of the wheel. Water from the millstock flowed into the wheel at a point on the side of the rotation, hitting the blades and imparting momentum on the blades. The weight of the water then pulls the wheel downward through gravity. The breastshot wheel thus extracts power from both the weight and momentum of the water. Breastshot wheels reached efficiencies of up to 50%, compared to typical undershot wheels around 30%.

During a widespread French effort to improve water wheel designs, Jean Charles de Borda discovered the efficiency was a function of the relative speed of the water entering and exiting the wheel. As the incoming speed is a function of the water source, the key to extracting more power from a water wheel was reducing the velocity of the exhaust water as much as possible. He also noticed that water entering the system was often slowed before it reached the wheel, through a variety of mechanisms. He suggested that great improvements could be made by improving these aspects of wheel design.

In a conventional breastshot design, the water would flow off the edge of the headrace, and fall onto the blades of the wheel. In comparison, the Sagebien wheel had the water enter through a channel, flowing horizontally. The blades of the wheel were angled so they entered the flowing water vertically. The inside of the wheel was open, so the water could flow up into the channel without the air pressure building up and impeding the flow. The water flowed back out again after a short time, the wheel turning perhaps 30 to 45 degrees, into the lower altitude tailrace.

Sagebien built his first wheel around 1850, and made the first full-sized version, at 6 to 7 horsepower, at a flour mill in Ronquerolles in 1851. Testing demonstrated this wheel operated at about 85% efficiency, well in advance of any design of the era. This attracted little notice at the time, in the era of the steam engine, but by 1857 he had 17 in operation and people started to take note. One test in December 1861 suggested an efficiency of one model of 103%, until a new gauge reduced this to a still-excellent 93%. By 1868, more than 60 Sagebien wheels were in use in northern France, and the French Academy of Sciences awarded him the Fourneyron Prix in 1875.

Water Turbine

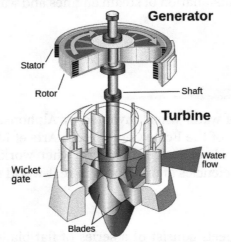

Kaplan turbine and electrical generator cut-away view.

The runner of the small water turbine

A water turbine is a rotary machine that converts kinetic energy and potential energy of water into mechanical work.

Water turbines were developed in the 19th century and were widely used for industrial power prior to electrical grids. Now they are mostly used for electric power generation. Water turbines are mostly found in dams to generate electric power from water kinetic energy.

History

The construction of a Ganz water Turbo Generator in Budapest in 1886

Water wheels have been used for hundreds of years for industrial power. Their main shortcoming is size, which limits the flow rate and head that can be harnessed. The migration from water wheels to modern turbines took about one hundred years. Development occurred during the Industrial revolution, using scientific principles and methods. They also made extensive use of new materials and manufacturing methods developed at the time.

Swirl

The word turbine was introduced by the French engineer Claude Burdin in the early 19th century and is derived from the Latin word for "whirling" or a "vortex". The main difference between early water turbines and water wheels is a swirl component of the water which passes energy to a spinning rotor. This additional component of motion allowed the turbine to be smaller than a water wheel of the same power. They could process more water by spinning faster and could harness much greater heads. (Later, impulse turbines were developed which didn't use swirl).

Timeline

Roman turbine mill at Chemtou, Tunisia. The tangential water inflow of the millrace made the submerged horizontal wheel in the shaft turn like a true turbine.

A Francis turbine runner, rated at nearly one million hp (750 MW), being installed at the Grand Coulee Dam, United States.

A propeller-type runner rated 28,000 hp (21 MW)

The earliest known water turbines date to the Roman Empire. Two helix-turbine mill sites of almost identical design were found at Chemtou and Testour, modern-day Tunisia, dating to the late 3rd or early 4th century AD. The horizontal water wheel with angled blades was installed at the bottom of a water-filled, circular shaft. The water from the mill-race entered the pit tangentially, creating a swirling water column which made the fully submerged wheel act like a true turbine.

Fausto Veranzio in his book Machinae Novae (1595) described a vertical axis mill with a rotor similar to that of a Francis turbine.

Johann Segner developed a reactive water turbine (Segner wheel) in the mid-18th century in Kingdom of Hungary. It had a horizontal axis and was a precursor to modern water turbines. It is a very simple machine that is still produced today for use in small hydro sites. Segner worked with Euler on some of the early mathematical theories of turbine design. In the 18th century, a Dr. Barker invented a similar reaction hydraulic turbine that became popular as a lecture-hall demonstration. The only known surviving example of this type of engine used in power production, dating from 1851, is found at Hacienda Buena Vista in Ponce, Puerto Rico.

In 1820, Jean-Victor Poncelet developed an inward-flow turbine.

In 1826, Benoît Fourneyron developed an outward-flow turbine. This was an efficient machine (~80%) that sent water through a runner with blades curved in one dimension. The stationary outlet also had curved guides.

In 1844, Uriah A. Boyden developed an outward flow turbine that improved on the performance of the Fourneyron turbine. Its runner shape was similar to that of a Francis turbine.

In 1849, James B. Francis improved the inward flow reaction turbine to over 90% efficiency. He also conducted sophisticated tests and developed engineering methods for water turbine design. The Francis turbine, named for him, is the first modern water turbine. It is still the most widely used water turbine in the world today. The Francis turbine is also called a radial flow turbine, since water flows from the outer circumference towards the centre of runner.

Inward flow water turbines have a better mechanical arrangement and all modern reaction water turbines are of this design. As the water swirls inward, it accelerates, and transfers energy to the runner. Water pressure decreases to atmospheric, or in some cases subatmospheric, as the water passes through the turbine blades and loses energy.

Around 1890, the modern fluid bearing was invented, now universally used to support heavy water turbine spindles. As of 2002, fluid bearings appear to have a mean time between failures of more than 1300 years.

Around 1913, Viktor Kaplan created the Kaplan turbine, a propeller-type machine. It was an evolution of the Francis turbine but revolutionized the ability to develop low-head hydro sites.

New Concept

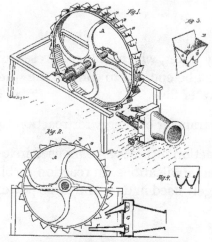

Figure from Pelton's original patent (October 1880)

All common water machines until the late 19th century (including water wheels) were basically reaction machines; water *pressure* head acted on the machine and produced work. A reaction turbine needs to fully contain the water during energy transfer.

In 1866, California millwright Samuel Knight invented a machine that took the impulse system to a new level. Inspired by the high pressure jet systems used in hydraulic mining in the gold fields, Knight developed a bucketed wheel which captured the energy of a free jet, which had converted a high head (hundreds of vertical feet in a pipe or penstock) of water to kinetic energy. This is called an impulse or tangential turbine. The water's velocity, roughly twice the velocity of the bucket periphery, does a u-turn in the bucket and drops out of the runner at low velocity.

In 1879, Lester Pelton, experimenting with a Knight Wheel, developed a Pelton wheel (double bucket design), which exhausted the water to the side, eliminating some energy loss of the Knight wheel which exhausted some water back against the center of the wheel. In about 1895, William Doble improved on Pelton's half-cylindrical bucket form with an elliptical bucket that included a cut in it to allow the jet a cleaner bucket entry. This is the modern form of the Pelton turbine which today achieves up to 92% efficiency. Pelton had been quite an effective promoter of his design and although Doble took over the Pelton company he did not change the name to Doble because it had brand name recognition.

Turgo and cross-flow turbines were later impulse designs.

Theory of Operation

Flowing water is directed on to the blades of a turbine runner, creating a force on the blades. Since the runner is spinning, the force acts through a distance (force acting through a distance is the definition of work). In this way, energy is transferred from the water flow to the turbine

Water turbines are divided into two groups; reaction turbines and impulse turbines.

The precise shape of water turbine blades is a function of the supply pressure of water, and the type of impeller selected.

Reaction Turbines

Reaction turbines are acted on by water, which changes pressure as it moves through the turbine and gives up its energy. They must be encased to contain the water pressure (or suction), or they must be fully submerged in the water flow.

Newton's third law describes the transfer of energy for reaction turbines.

Most water turbines in use are reaction turbines and are used in low (<30 m or 100 ft) and medium (30–300 m or 100–1,000 ft) head applications. In reaction turbine pressure drop occurs in both fixed and moving blades. It is largely used in dam and large power plants

Impulse Turbines

Impulse turbines change the velocity of a water jet. The jet pushes on the turbine's curved blades which changes the direction of the flow. The resulting change in momentum (impulse) causes a force on the turbine blades. Since the turbine is spinning, the force acts through a distance (work) and the diverted water flow is left with diminished energy. An impulse turbine is one which the pressure of the fluid flowing over the rotor blades is constant and all the work output is due to the change in kinetic energy of the fluid.

Prior to hitting the turbine blades, the water's pressure (potential energy) is converted to kinetic energy by a nozzle and focused on the turbine. No pressure change occurs at the turbine blades, and the turbine doesn't require a housing for operation.

Newton's second law describes the transfer of energy for impulse turbines.

Impulse turbines are often used in very high (>300m/1000 ft) head applications.

Power

The power available in a stream of water is;

$$P = \eta \cdot \rho \cdot g \cdot h \cdot \dot{q}$$

where:

- P = power (J/s or watts)

- η = turbine efficiency

- ρ = density of water (kg/m³)

- g = acceleration of gravity (9.81 m/s²)

- h = head (m). For still water, this is the difference in height between the inlet and outlet surfaces. Moving water has an additional component added to account for the kinetic energy of the flow. The total head equals the *pressure head* plus *velocity head*.

- $\dot{q}$ = flow rate (m³/s)

Pumped-storage Hydroelectricity

Some water turbines are designed for pumped-storage hydroelectricity. They can reverse flow and operate as a pump to fill a high reservoir during off-peak electrical hours, and then revert to a water turbine for power generation during peak electrical demand. This type of turbine is usually a Deriaz or Francis turbine in design.

Efficiency

Large modern water turbines operate at mechanical efficiencies greater than 90%.

Types of Water Turbines

Various types of water turbine runners. From left to right: Pelton wheel, two types of Francis turbine and Kaplan turbine.

Reaction turbines:

- VLH turbine

- Francis turbine

- Kaplan turbine

- Tyson turbine

- Gorlov helical turbine

Impulse turbine

- Water wheel

- Pelton wheel

- Turgo turbine

- Cross-flow turbine (also known as the Bánki-Michell turbine, or Ossberger turbine)

- Jonval turbine

- Reverse overshot water-wheel

- Screw turbine

- Barkh Turbine

Design and Application

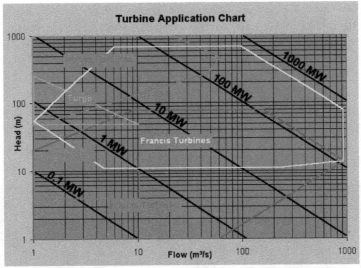

Turbine selection is based on the available water head, and less so on the available flow rate. In general, impulse turbines are used for high head sites, and reaction turbines are used for low head sites. Kaplan turbines with adjustable blade pitch are well-adapted to wide ranges of flow or head conditions, since their peak efficiency can be achieved over a wide range of flow conditions.

Small turbines (mostly under 10 MW) may have horizontal shafts, and even fairly large bulb-type turbines up to 100 MW or so may be horizontal. Very large Francis and Kaplan machines usually have vertical shafts because this makes best use of the available head, and makes installation of a generator more economical. Pelton wheels may be either vertical or horizontal shaft machines because the size of the machine is so much less than the available head. Some impulse turbines use multiple jets per runner to balance shaft thrust. This also allows for the use of a smaller turbine runner, which can decrease costs and mechanical losses.

Typical Range of Heads

• Water wheel	0.2 < H < 4 (H = head in m)
• Screw turbine	1 < H < 10
• VLH turbine	1.5 < H < 4.5
• Kaplan turbine	20 < H < 40
• Francis turbine	40 < H < 600
• Pelton wheel	50 < H < 1300
• Turgo turbine	50 < H < 250

Specific Speed

The specific speed n_s of a turbine characterizes the turbine's shape in a way that is not related to its size. This allows a new turbine design to be scaled from an existing design of known performance. The specific speed is also the main criteria for matching a specific hydro site with the correct turbine type. The specific speed is the speed with which the turbine turns for a particular discharge Q, with unit head and thereby is able to produce unit power.

Affinity Laws

Affinity laws allow the output of a turbine to be predicted based on model tests. A miniature replica of a proposed design, about one foot (0.3 m) in diameter, can be tested and the laboratory measurements applied to the final application with high confidence. Affinity laws are derived by requiring similitude between the test model and the application.

Flow through the turbine is controlled either by a large valve or by wicket gates arranged around the outside of the turbine runner. Differential head and flow can be plotted for a number of different values of gate opening, producing a hill diagram used to show the efficiency of the turbine at varying conditions.

Runaway Speed

The runaway speed of a water turbine is its speed at full flow, and no shaft load. The turbine will be designed to survive the mechanical forces of this speed. The manufacturer will supply the runaway speed rating.

Control Systems

Operation of a flyball governor to control speeds of a water turbine

Different designs of governors have been used since the mid-19th century to control the speeds of the water turbines. A variety of flyball systems, or first-generation governors, were used during the first 100 years of water turbine speed controls. In early flyball systems, the flyball component countered by a spring acted directly to the valve of the turbine or the wicket gate to control the amount of water that enters the turbines. Newer systems with mechanical governors started around 1880. An early mechanical governors is a servomechanism that comprises a series of gears that use the turbine's speed to drive the flyball and turbine's power to drive the control mechanism. The mechanical governors were continued to be enhanced in power amplification through the use of gears and the dynamic behavior. By 1930, the mechanical governors had many parameters that could be set on the feedback system for precise controls. In the later part of the twentieth century, electronic governors and digital systems started to replace the mechanical governors. In the electronic governors, also known as second-generation governors, the flyball was replaced by rotational speed sensor but the controls were still done through analog systems. In the modern systems, also known as third-generation governors, the controls are performed digitally by algorithms that are programmed to the computer of the governor.

Turbine Blade Materials

Given that the turbine blades in a water turbine are constantly exposed to water and dynamic forces, they need to have high corrosion resistance and strength. The most common material used in overlays on carbon steel runners in water turbines are austenitic steel alloys that have 17% to 20% chromium to increase stability of the film which improves aqueous corrosion resistance. The chromium content in these steel alloys exceed the minimum of 12% chromium required to exhibit some atmospheric corrosion resistance. Having a higher chromium concentration in the steel alloys allows for a much longer lifespan of the turbine blades. Currently, the blades are made of martensitic stainless steels which have high strength compared to austenitic stainless steels by a factor of 2. Besides corrosion resistance and strength as the criteria for material selection, weld-ability and density of the turbine blade. Greater weld-ability allows for easier repair of the turbine blades. This also allows for higher weld quality which results in a better repair. Selecting a material with low density is important to achieve higher efficiency because the lighter blades rotate more easily. The most common material used in Kaplan Turbine blades are stainless steel alloys (SS). The different alloys used are SS(16Cr-5Ni), SS(13Cr-4Ni), SS(13Cr-1Ni). The martensitic stainless steel alloys have high strength, thinner sections than standard carbon steel, and reduced mass that enhances the hydrodynamic flow conditions and efficiency of the water turbine. The SS(13Cr-4Ni) has been shown to have improved erosion resistance at all angles of attack through the process of laser hardening. It is important to minimize erosion in order to maintain high efficiencies because erosion negatively impacts the hydraulic profile of the blades which reduces the relative ease to rotate.

Maintenance

Turbines are designed to run for decades with very little maintenance of the main elements; overhaul intervals are on the order of several years. Maintenance of the runners and parts exposed to water include removal, inspection, and repair of worn parts.

Normal wear and tear includes pitting corrosion from cavitation, fatigue cracking, and abrasion from suspended solids in the water. Steel elements are repaired by welding, usually with stainless steel rods. Damaged areas are cut or ground out, then welded back up to their original or an improved profile. Old turbine runners may have a significant amount of stainless steel added this way by the end of their lifetime. Elaborate welding procedures may be used to achieve the highest quality repairs.

A Francis turbine at the end of its life showing pitting corrosion, fatigue cracking and a catastrophic failure. Earlier repair jobs that used stainless steel weld rods are visible.

Other elements requiring inspection and repair during overhauls include bearings, packing box and shaft sleeves, servomotors, cooling systems for the bearings and generator coils, seal rings, wicket gate linkage elements and all surfaces.

Environmental Impact

Water turbines are generally considered a clean power producer, as the turbine causes essentially no change to the water. They use a renewable energy source and are designed to operate for decades. They produce significant amounts of the world's electrical supply.

Historically there have also been negative consequences, mostly associated with the dams normally required for power production. Dams alter the natural ecology of rivers, potentially killing fish, stopping migrations, and disrupting peoples' livelihoods. For example, American Indian tribes in the Pacific Northwest had livelihoods built around salmon fishing, but aggressive dam-building destroyed their way of life. Dams also cause less obvious, but potentially serious consequences, including increased evaporation of water (especially in arid regions), buildup of silt behind the dam, and changes to water temperature and flow patterns. In the United States, it is now illegal to block the migration of fish, for example the white sturgeon in North America, so fish ladders must be provided by dam builders.

Francis Turbine

The Francis turbine is a type of water turbine that was developed by James B. Francis in Lowell, Massachusetts. It is an inward-flow reaction turbine that combines radial and axial flow concepts.

Francis turbines are the most common water turbine in use today. They operate in a water head from 40 to 600 m (130 to 2,000 ft) and are primarily used for electrical power production. The electric generators which most often use this type of turbine have a power output which generally ranges just a few kilowatts up to 800 MW, though mini-hydro installations may be lower. Penstock (input pipes) diameters are between 3 and 33 feet (0.91 and 10.06 metres). The speed range of the turbine is from 75 to 1000 rpm. Wicket gates around the outside of the turbine's rotating runner control the rate of water flow through the turbine for different power production rates. Francis turbines are almost always mounted with the shaft vertical to isolate water from the generator. This also facilitates installation and maintenance.

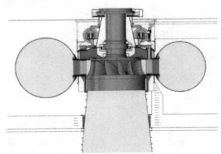

Side-view cutaway of a vertical Francis turbine. Here water enters horizontally in a spiral shaped pipe (spiral case) wrapped around the outside of the turbine's rotating *runner* and exits vertically down through the center of the turbine.

Development

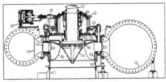

Figure 5 - Vertical Francis Turbine

PARTS LIST FOR FRANCIS AND KAPLAN TURBINE DRAWINGS			
1	Turbine Runner	7	Draft Tube
1a	Runner Cone	8	Discharge Ring
1b	Runner Crown (Francis)	9	Turbine Shaft
1c	Runner Band (Francis)	10	Turbine Guide Bearing
1d	Runner Bucket (Francis)	11	Wicket Gate Servomotors
1e	Runner Hub (Kaplan)	12	Servomotor Connecting Ring
1f	Runner Blade (Kaplan)	13	Wicket Gate Operating Ring or Shift Ring
2	Wearing Rings or Seal Rings (Francis)	14	Wicket Gate Link
3	Facing Plates or Curb Plates	15	Wicket Gate Arm
4	Spiral Case or Scroll Case	16	Packing Box or Stuffing Box (Mechanical Seals)
5	Stay Vane	17	Head Cover
6	Wicket Gate	18	Runner Blade Servomotor (Kaplan)

Francis turbine parts

Pawtucket Gatehouse in Lowell, Massachusetts; site of the first Francis turbine

Water wheels of different types have been used historically for over 1000 years to power mills of all types, but they were relatively inefficient. Nineteenth-century efficiency improvements of water turbines allowed them to replace nearly all water wheel applications and compete with steam engines wherever water power was available. After electric generators were developed in the late 1800s turbines were a natural source of generator power where potential hydro-power sources existed.

Francis Runner, Grand Coulee Dam

In 1826 Benoit Fourneyron developed a high efficiency (80%) outward-flow water turbine. Water was directed tangentially through the turbine runner, causing it to spin. Jean-Victor Poncelet designed an inward-flow turbine in about 1820 that used the same principles. S. B. Howd obtained a US patent in 1838 for a similar design.

In 1848 James B. Francis, while working as head engineer of the Locks and Canals company in the water wheel-powered textile factory city of Lowell, Massachusetts, improved on these designs to create more efficient turbines. He applied scientific principles and testing methods to produce a very efficient turbine design. More importantly, his mathematical and graphical calculation methods improved turbine design and engineering. His analytical methods allowed confident design of high efficiency turbines to precisely match a site's water flow and pressure (water head).

Components

A Francis turbine consists of the following main parts:

Spiral casing: The spiral casing around the runner of the turbine is known as the volute casing or scroll case. Throughout its length, it has numerous openings at regular intervals to allow the working fluid to impinge on the blades of the runner. These openings convert the pressure energy of the fluid into momentum energy just before the fluid impinges on the blades. This maintains a constant flow rate despite the fact that numerous openings have been provided for the fluid to enter the blades, as the cross-sectional area of this casing decreases uniformly along the circumference.

Guide or stay vanes: The primary function of the guide or stay vanes is to convert the pressure energy of the fluid into the momentum energy. It also serves to direct the flow at design angles to the runner blades.

Runner blades:Runner blades are the heart of any turbine. These are the centers where the fluid strikes and the tangential force of the impact causes the shaft of the turbine to rotate, producing torque. Close attention in design of blade angles at inlet and outlet is necessary, as these are major parameters affecting power production.

Draft tube: The draft tube is a conduit which connects the runner exit to the tail race where the water is discharged from the turbine. Its primary function is to reduce the velocity of discharged water to minimize the loss of kinetic energy at the outlet. This permits the turbine to be set above the tail water without appreciable drop of available head.

Theory of Operation

Three Gorges Dam Francis turbine runner

The Francis turbine is a type of reaction turbine, a category of turbine in which the working fluid comes to the turbine under immense pressure and the energy is extracted by the turbine blades from the working fluid. A part of the energy is given up by the fluid because of pressure changes occurring in the blades of the turbine, quantified by the expression of Degree of reaction, while the remaining part of the energy is extracted by the volute casing of the turbine. At the exit, water acts on the spinning cup-shaped runner features, leaving at low velocity and low swirl with very little kinetic or potential energy left. The turbine's exit tube is shaped to help decelerate the water flow and recover the pressure.

Francis Turbine (exterior view) attached to a generator

Cut-away view, with wicket gates (yellow) at minimum flow setting

Cut-away view, with wicket gates (yellow) at full flow setting

Blade Efficiency

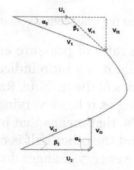

Ideal velocity diagram, illustrating that in ideal cases the whirl component of outlet velocity is zero and the flow is completely axial

Usually the flow velocity (velocity perpendicular to the tangential direction) remains constant throughout, i.e. $V_{f1} = V_{f2}$ and is equal to that at the inlet to the draft tube. Using Euler turbine equation, $E/m = e = V_{w1}U_1$, where e is the energy transfer to the rotor per unit mass of the fluid. From the inlet velocity triangle,

$$V_{w1} = V_{f1}\cot\alpha_1$$

and

$$U_1 = V_{f1}(\cot\alpha_1 + \cot\beta_1),$$

Therefore

$$e = V_{f1}^2\cot\alpha_1(\cot\alpha_1 + \cot\beta_1).$$

The loss of kinetic energy per unit mass becomes $V_{f2}^2/2$.

Therefore, neglecting friction, the blade efficiency becomes

$$\eta_b = \frac{e}{(e + V_{f2}^2/2)},$$

i.e.

$$\eta_b = \frac{2V_{f1}^2(\cot\alpha_1(\cot\alpha_1+\cot\beta_1))}{V_{f2}^2+2V_{f1}^2(\cot\alpha_1(\cot\alpha_1+\cot\beta_1))}.$$

Degree of Reaction

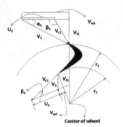

Actual velocity diagram, illustrating that the whirl component of the outlet velocity is non-zero

Degree of reaction can be defined as the ratio of pressure energy change in the blades to total energy change of the fluid. This means that it is a ratio indicating the fraction of total change in fluid pressure energy occurring in the blades of the turbine. Rest of the changes occur in the stator blades of the turbines and the volute casing as it has a varying cross-sectional area. For example, if the degree of reaction is given to be 50%, that means that half of the total energy change of the fluid is taking place in the rotor blades and the other half is occurring in the stator blades. If the degree of reaction is zero it means that the energy changes due to the rotor blades is zero, leading to a different turbine design called Pelton Turbine.

R=e-1/2(V_1^2-V_{f2}^2)/e

Now, putting the value of 'e' from above and using (V_1^2-V_{f2}^2=V_{f1}^2cotα_2 as V_{f2}=V_{f1})

R=1-(cotα_1/2(cotα_1+cotβ_1))

Application

Francis Inlet Scroll, Grand Coulee Dam

Francis turbines may be designed for a wide range of heads and flows. This, along with their high efficiency, has made them the most widely used turbine in the world. Francis type units cover a head range from 40 to 600 m (130 to 2,000 ft), and their connected generator output power varies from just a few kilowatts up to 800 MW. Large Francis turbines are individually designed for each site to operate with the given water supply and water head at the highest possible efficiency, typically over 90%.

Small Swiss-made Francis turbine

In contrast to the Pelton turbine, the Francis turbine operates at its best completely filled with water at all times. The turbine and the outlet channel may be placed lower than the lake or sea level outside, reducing the tendency for cavitation.

In addition to electrical production, they may also be used for pumped storage, where a reservoir is filled by the turbine (acting as a pump) driven by the generator acting as a large electrical motor during periods of low power demand, and then reversed and used to generate power during peak demand. These pump storage reservoirs act as large energy storage sources to store "excess" electrical energy in the form of water in elevated reservoirs. This is one of a few methods that temporary excess electrical capacity can be stored for later utilization.

Kaplan Turbine

A Bonneville Dam Kaplan turbine after 61 years of service

The Kaplan turbine is a propeller-type water turbine which has adjustable blades. It was developed in 1913 by Austrian professor Viktor Kaplan, who combined automatically adjusted propeller blades with automatically adjusted wicket gates to achieve efficiency over a wide range of flow and water level.

The Kaplan turbine was an evolution of the Francis turbine. Its invention allowed efficient power production in low-head applications that was not possible with Francis turbines. The head ranges from 10–70 metres and the output from 5 to 200 MW. Runner diameters are between 2 and 11 metres. Turbines rotate at a constant rate, which varies from facility to facility. That rate ranges from as low as 69.2 rpm (Bonneville North Powerhouse, Washington U.S.) to 429 rpm. The Kaplan turbine installation believed to generate the most power from its nominal head of 34.65 m is as of 2013 the Tocoma Dam Power Plant (Venezuela) Kaplan turbine generating 235 MW with each of ten 4.8 m diameter runners.

Kaplan turbines are now widely used throughout world in high-flow, low-head power production.

On this Kaplan runner the pivots at the base of the blade are visible; these allow the angle of the blades to be changed while running. The hub contains hydraulic cylinders for adjusting the angle.

Development

Viktor Kaplan living in Brno, Czech Republic, obtained his first patent for an adjustable blade propeller turbine in 1912. But the development of a commercially successful machine would take another decade. Kaplan struggled with cavitation problems, and in 1922 abandoned his research for health reasons.

In 1919 Kaplan installed a demonstration unit at Poděbrady, Czechoslovakia. In 1922 Voith introduced an 1100 HP (about 800 kW) Kaplan turbine for use mainly on rivers. In 1924 an 8 MW unit went on line at Lilla Edet, Sweden. This marked the commercial success and widespread acceptance of Kaplan turbines.

Theory of Operation

The Kaplan turbine is an outward flow reaction turbine, which means that the working fluid changes pressure as it moves through the turbine and gives up its energy. Power is recovered from both the hydrostatic head and from the kinetic energy of the flowing water. The design combines features of radial and axial turbines.

The inlet is a scroll-shaped tube that wraps around the turbine's wicket gate. Water is directed tangentially through the wicket gate and spirals on to a propeller shaped runner, causing it to spin.

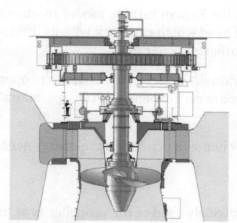

Vertical Kaplan Turbine (courtesy Voith-Siemens).

The outlet is a specially shaped draft tube that helps decelerate the water and recover kinetic energy.

The turbine does not need to be at the lowest point of water flow as long as the draft tube remains full of water. A higher turbine location, however, increases the suction that is imparted on the turbine blades by the draft tube. The resulting pressure drop may lead to cavitation.

Variable geometry of the wicket gate and turbine blades allow efficient operation for a range of flow conditions. Kaplan turbine efficiencies are typically over 90%, but may be lower in very low head applications.

Current areas of research include CFD driven efficiency improvements and new designs that raise survival rates of fish passing through.

Because the propeller blades are rotated on high-pressure hydraulic oil bearings, a critical element of Kaplan design is to maintain a positive seal to prevent emission of oil into the waterway. Discharge of oil into rivers is not desirable because of the waste of resources and resulting ecological damage.

Applications

Viktor Kaplan Turbine Technisches Museum Wien

Kaplan turbines are widely used throughout the world for electrical power production. They cover the lowest head hydro sites and are especially suited for high flow conditions.

Inexpensive micro turbines on the Kaplan turbine model are manufactured for individual power production designed for 3 m of head which can work with as little as 0.3 m of head at a highly reduced performance provided sufficient water flow.

Large Kaplan turbines are individually designed for each site to operate at the highest possible efficiency, typically over 90%. They are very expensive to design, manufacture and install, but operate for decades.

They have recently found a new home in offshore wave energy generation.

Variations

The Kaplan turbine is the most widely used of the propeller-type turbines, but several other variations exist:

- *Propeller turbines* have non-adjustable propeller vanes. They are used in where the range of flow / power is not large. Commercial products exist for producing several hundred watts from only a few feet of head. Larger propeller turbines produce more than 100 MW. At the La Grande-1 generating station in northern Quebec, 12 propeller turbines generate 1368 MW.

- *Bulb* or *tubular turbines* are designed into the water delivery tube. A large bulb is centered in the water pipe which holds the generator, wicket gate and runner. Tubular turbines are a fully axial design, whereas Kaplan turbines have a radial wicket gate.

- *Pit turbines* are bulb turbines with a gear box. This allows for a smaller generator and bulb.

- *Straflo turbines* are axial turbines with the generator outside of the water channel, connected to the periphery of the runner.

- *S-turbines* eliminate the need for a bulb housing by placing the generator outside of the water channel. This is accomplished with a jog in the water channel and a shaft connecting the runner and generator.

- The *VLH turbine* an open flow, very low head "kaplan" turbine slanted at an angle to the water flow. It has a large diameter >3.55m, is low speed using a directly connected shaft mounted permanent magnet alternator with electronic power regulation and is very fish friendly (<5% mortality).

- *Tyson turbines* are a fixed propeller turbine designed to be immersed in a fast flowing river, either permanently anchored in the river bed, or attached to a boat or barge.

model of a bulb or tubular turbines

model of a S-turbine

Cross-flow Turbine

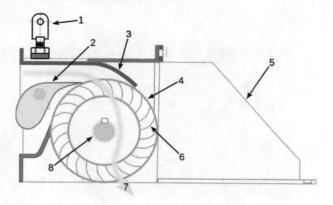

Diagram of a Cross-flow turbine
1 — air-venting valve
2 — distributor
3 — turbine casing (all thick grey)
4 — runner
5 — removable rear casing
6 — blades
7 — water flow
8 — shaft

A cross-flow turbine, Bánki-Michell turbine, or Ossberger turbine is a water turbine developed by the Australian Anthony Michell, the Hungarian Donát Bánki and the German Fritz Ossberger. Michell obtained patents for his turbine design in 1903, and the manufacturing company Weymouth made it for many years. Ossberger's first patent was granted in 1933 ("Free Jet Turbine" 1922, Imperial Patent No. 361593 and the "Cross Flow Turbine" 1933, Imperial Patent No. 615445), and he manufactured this turbine as a standard product. Today, the company founded by Ossberger is the leading manufacturer of this type of turbine.

Unlike most water turbines, which have axial or radial flows, in a cross-flow turbine the water passes through the turbine transversely, or across the turbine blades. As with a water wheel, the water is admitted at the turbine's edge. After passing to the inside of the runner, it leaves on the opposite side, going outward. Passing through the runner twice provides additional efficiency.

When the water leaves the runner, it also helps clean it of small debris and pollution. The cross-flow turbine is a low-speed machine that is well suited for locations with a low head but high flow.

Although the illustration shows one nozzle for simplicity, most practical cross-flow turbines have two, arranged so that the water flows do not interfere.

Cross-flow turbines are often constructed as two turbines of different capacity that share the same shaft. The turbine wheels are the same diameter, but different lengths to handle different volumes at the same pressure. The subdivided wheels are usually built with volumes in ratios of 1:2. The subdivided regulating unit, the guide vane system in the turbine's upstream section, provides flexible operation, with 33, 66 or 100% output, depending on the flow. Low operating costs are obtained with the turbine's relatively simple construction.

Details of Design

Ossberger turbine section

The turbine consists of a cylindrical water wheel or runner with a horizontal shaft, composed of numerous blades (up to 37), arranged radially and tangentially. The blade's edges are sharpened to reduce resistance to the flow of water. A blade is made in a part-circular cross-section (pipe cut over its whole length). The ends of the blades are welded to disks to form a cage like a hamster cage and are sometimes called "squirrel cage turbines"; instead of the bars, the turbine has the trough-shaped steel blades.

The water flows first from the outside of the turbine to its inside. The regulating unit, shaped like a vane or tongue, varies the cross-section of the flow. The water jet is directed towards the cylindrical runner by nozzle. The water enters the runner at an angle of about 45/120 degrees, transmitting some of the water's kinetic energy to the active cylindrical blades.

Ossberger turbine runner

The regulating device controls the flow based on the power needed, and the available water. The ratio is that (0–100%) of the water is admitted to 0-100%×30/4 blades. Water admission to the two nozzles is throttled by two shaped guide vanes. These divide and direct the flow so that the water enters the runner smoothly for any width of opening. The guide vanes should seal to the edges of the turbine casing so that when the water is low, they can shut off the water supply. The guide vanes therefore act as the valves between the penstock and turbine. Both guide vanes can be set by control levers, to which an automatic or manual control may be connected.

The turbine geometry (nozzle-runner-shaft) assures that the water jet is effective. The water acts on the runner twice, but most of the power is transferred on the first pass, when the water enters the runner. Only ⅓ of the power is transferred to the runner when the water is leaving the turbine.

The water flows through the blade channels in two directions: outside to inside, and inside to outside. Most turbines are run with two jets, arranged so two water jets in the runner will not affect each other. It is, however, essential that the turbine, head and turbine speed are harmonised.

The cross-flow turbine is of the impulse type, so the pressure remains constant at the runner.

Advantages

The peak efficiency of a cross-flow turbine is somewhat less than a Kaplan, Francis or Pelton turbine. However, the cross-flow turbine has a flat efficiency curve under varying load. With a split runner and turbine chamber, the turbine maintains its efficiency while the flow and load vary from 1/6 to the maximum.

Since it has a low price, and good regulation, cross-flow turbines are mostly used in mini and micro hydropower units of less than two thousand kW and with heads less than 200 m.

Particularly with small run-of-the-river plants, the flat efficiency curve yields better annual performance than other turbine systems, as small rivers' water is usually lower in some months. The efficiency of a turbine determines whether electricity is produced during the periods when rivers have low flows. If the turbines used have high peak efficiencies, but behave poorly at partial load, less annual performance is obtained than with turbines that have a flat efficiency curve.

Due to its excellent behaviour with partial loads, the cross-flow turbine is well-suited to unattended electricity production. Its simple construction makes it easier to maintain than other turbine types; only two bearings must be maintained, and there are only three rotating elements. The mechanical system is simple, so repairs can be performed by local mechanics.

Another advantage is that it can often clean itself. As the water leaves the runner, leaves, grass etc. will not remain in the runner, preventing losses. Therefore, although the turbine's efficiency is somewhat lower, it is more reliable than other types. No runner cleaning is normally necessary, e.g. by flow inversion or variations of the speed. Other turbine types are clogged more easily, and consequently face power losses despite higher nominal efficiencies.

Gorlov Helical Turbine

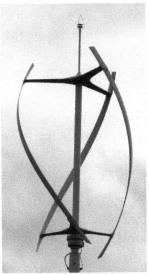

quietrevolution QR5 wind turbine

The Gorlov helical turbine (GHT) is a water turbine evolved from the Darrieus turbine design by altering it to have helical blades/foils. It was patented in a series of patents from September 19, 1995 to July 3, 2001 and won 2001 ASME Thomas A. Edison Patent Award. GHT was invented by Professor Alexander M. Gorlov of Northeastern University.

A.M.Gorlov with his turbine.

The physical principles of the GHT work are the same as for its main prototype, the Darrieus turbine, and for the family of similar Vertical axis wind turbines which includes also Turby wind turbine Quietrevolution wind turbine Urban Green Energy. GHT, turby and quietrevolution solved pulsatory torque issues by using the helical twist of the blades.

Fluid Performance

The term "foil" is used to describe the shape of the blade cross-section at a given point, with no distinction for the type of fluid, (thus referring to either an "airfoil" or "hydrofoil"). In the helical design, the blades curve around the axis, which has the effect of evenly distributing the foil sections throughout the rotation cycle, so there is always a foil section at every possible angle of attack. In this way, the sum of the lift and drag forces on each blade do not change abruptly with rotation angle. The turbine generates a smoother torque curve, so there is much less vibration and noise than in the Darrieus design. It also minimizes peak stresses in the structure and materials, and facilitates self-starting of the turbine. In testing environments the GHT has been observed to have up to 35% efficiency in energy capture reported by several groups. "Among the other vertical-axis turbine systems, the Davis Hydro turbine, the EnCurrent turbine, and the Gorlov Helical turbine have all undergone scale testing at laboratory or sea. Overall, these technologies represent the current norm of tidal current development."

Turbine Axis-orientation

The main difference between the Gorlov helical turbine and conventional turbines is the orientation of the axis in relation to current flow. The GHT is a vertical-axis turbine which means the axis is positioned perpendicular to current flow, whereas traditional turbines are horizontal-axis turbines which means the axis is positioned parallel to the flow of the current. Fluid flows, such as wind, will naturally change direction, however they will still remain parallel to the ground. So in all vertical-axis turbines, the flow remains perpendicular to the axis, regardless of the flow direction, and the turbines always rotate in the same direction. This is one of the main advantages of vertical-axis turbines.

If the direction of the water flow is fixed, then the Gorlov turbine axis could be vertical or horizontal, the only requirement is orthogonality to the flow.

Airfoil / Hydrofoil

The GHT operates under a lift-based concept. The foil sections on the GHT are symmetrical, both top-to-bottom and also from the leading-to-trailing edge. The GHT can actually spin equally well in either direction. The GHT works under the same principle as the Darrieus turbine; that is, it relies upon the movement of the foils in order to change the apparent direction of the flow relative to the foils, and thus change the (apparent) "angle of attack" of the foil.

Environmental Issues

A GHT is proposed for low-head micro hydro installations, when construction of a dam is undesirable. The GHT is an example of damless hydro technology. The technology may potentially offer cost and environmental benefits over dam-based micro-hydro systems.

Some advantages of damless hydro are that it eliminates the potential for failure of a dam, which improves public safety. It also eliminates the initial cost of dam engineering, construction and maintenance, reduces the environmental and ecological complications, and potentially simplifies the regulatory issues put into law specifically to mitigate the problems with dams.

In general, a major ecological issue with hydropower installations is their actual and perceived risk to aquatic life. It is claimed that a GHT spins slowly enough that fish can see it soon enough to swim around it. From preliminary tests in 2001, it was claimed that if a fish swims between the slowly moving turbine blades, the fish will not be harmed. Also it would be difficult for a fish to become lodged or stuck in the turbine, because the open spaces between the blades are larger than even the largest fish living in a small river. A fish also would not be tumbled around in a vortex, because the GHT does not create a lot of turbulence, so small objects would be harmlessly swept through with the current.

How it Works

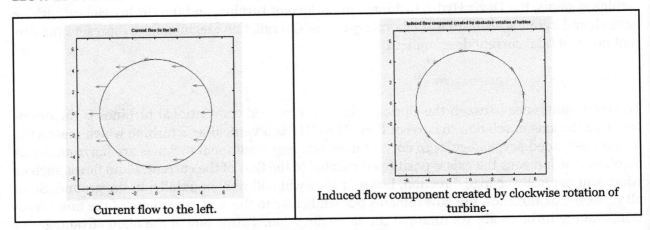

Current flow to the left.

Induced flow component created by clockwise rotation of turbine.

In this example the direction of the fluid flow is to the left.

As the turbine rotates, in this case in a clockwise direction, the motion of the foil through the fluid changes the apparent velocity and angle of attack (speed and direction) of the fluid with respect to the frame of reference of the foil. The combined effect of these two flow components (i.e. the vector sum), yields the net total "Apparent flow velocity" as shown in the next figure.

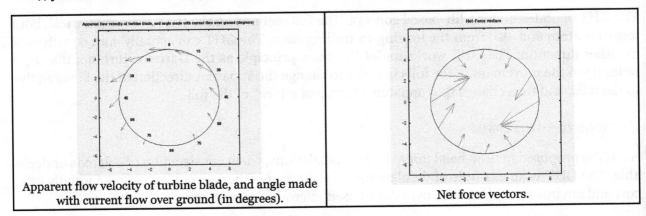

Apparent flow velocity of turbine blade, and angle made with current flow over ground (in degrees).

Net force vectors.

The action of this apparent flow on each foil section generates both a lift and drag force, the sum of which is shown in the figure above titled "Net force vectors". Each of these net force vectors can be split into two orthogonal vectors: a radial component and a tangential component, shown here as "Normal force" and "Axial force" respectively. The normal forces are opposed by the rigidity of the turbine structure and do not impart any rotational force or energy to the turbine. The remaining force component propels the turbine in the clockwise direction, and it is from this torque that energy can be harvested.

[With regards to the figure above left "Apparent flow velocity…", Lucid Energy Technologies, rights holder to the patent to the Gorlov Helical Turbine, notes that this diagram, with no apparent velocity at an azimuth angle of 180 degrees (blade at its point in rotation where it is instantaneously moving in downstream direction), may be subject to misinterpretation. This is because a zero apparently flow velocity could occur only at a tip speed ratio of unity (i.e. TSR=1, where the current flow induced by rotation equals the current flow). The GHT generally operates at a TSR substantially greater than unity.]

(The diagrams "Net Force Vectors" and "Normal Force Vectors" are partially incorrect. The downwind segments should show the vectors outside the circles. Otherwise there would be no net sideways loading on the turbine.) M Koester 2015.

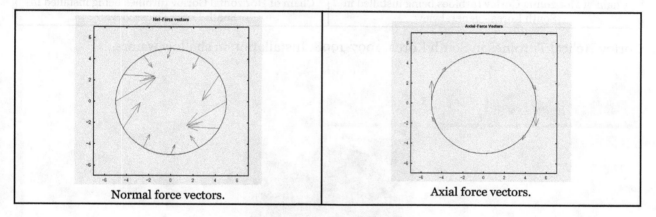

Normal force vectors.   Axial force vectors.

Commercial Use

Helical turbines in water stream generate mechanical power independent on direction of the water flow. Then electric generators assembled upon the common shaft transfer the power into electricity for the commercial use.

Chain of Horizontal Gorlov Turbines. TideGen by Ocean Renewable Power Company – possibility for shallow waters. (US Patents 5,451,137, Sep. 1995 and 5,642,984, Jul. 1997)

Chain of Horizontal Gorlov turbines being installed in Cobscook Bay, Maine, USA

Tidal Power Station with Gorlov Helical Turbines before deployment in the ocean. USA, Cobscook Bay, Maine, September, 2012.

Chain of Horizontal Gorlov turbines being installed in South Korea - general view

Chain of Horizontal Gorlov turbines being installed in South Korea - close view

Gorlov Helical Turbines in South Korea, 1997-1998. Installation in shallow waters.

Pelton Wheel

Old Pelton wheel from Walchensee Hydroelectric Power Station, Germany.

Assembly of a Pelton wheel at Walchensee Hydroelectric Power Station, Germany.

The Pelton wheel is an impulse type water turbine. It was invented by Lester Allan Pelton in the

1870s. The Pelton wheel extracts energy from the impulse of moving water, as opposed to water's dead weight like the traditional overshot water wheel. Many variations of impulse turbines existed prior to Pelton's design, but they were less efficient than Pelton's design. Water leaving those wheels typically still had high speed, carrying away much of the dynamic energy brought to the wheels. Pelton's paddle geometry was designed so that when the rim ran at half the speed of the water jet, the water left the wheel with very little speed; thus his design extracted almost all of the water's impulse energy—which allowed for a very efficient turbine.

Function

Nozzles direct forceful, high-speed streams of water against a rotary series of spoon-shaped buckets, also known as impulse blades, which are mounted around the circumferential rim of a drive wheel—also called a runner. As the water jet impinges upon the contoured bucket-blades, the direction of water velocity is changed to follow the contours of the bucket. Water impulse energy exerts torque on the bucket-and-wheel system, spinning the wheel; the water stream itself does a "u-turn" and exits at the outer sides of the bucket, decelerated to a low velocity. In the process, the water jet's momentum is transferred to the wheel and hence to a turbine. Thus, "impulse" energy does work on the turbine. For maximum power and efficiency, the wheel and turbine system is designed such that the water jet velocity is twice the velocity of the rotating buckets. A very small percentage of the water jet's original kinetic energy will remain in the water, which causes the bucket to be emptied at the same rate it is filled, and thereby allows the high-pressure input flow to continue uninterrupted and without waste of energy. Typically two buckets are mounted side-by-side on the wheel, which permits splitting the water jet into two equal streams. This balances the side-load forces on the wheel and helps to ensure smooth, efficient transfer of momentum of the fluid jet of water to the turbine wheel.

Because water and most liquids are nearly incompressible, almost all of the available energy is extracted in the first stage of the hydraulic turbine. Therefore, Pelton wheels have only one turbine stage, unlike gas turbines that operate with compressible fluid. It is used for generating electricity.

Applications

Pelton wheels are the preferred turbine for hydro-power, when the available water source has relatively high hydraulic head at low flow rates, where the Pelton wheel geometry is most suitable. Pelton wheels are made in all sizes. There exist multi-ton Pelton wheels mounted on vertical oil pad bearings in hydroelectric plants. The largest units can be over 400 megawatts. The smallest Pelton wheels are only a few inches across, and can be used to tap power from mountain streams having flows of a few gallons per minute. Some of these systems use household plumbing fixtures for water delivery. These small units are recommended for use with 30 metres (100 ft) or more of head, in order to generate significant power levels. Depending on water flow and design, Pelton wheels operate best with heads from 15–1,800 metres (50–5,910 ft), although there is no theoretical limit.

Design Rules

The specific speed η_s parameter is independent of a particular turbine's

Compared to other turbine designs, the relatively low specific speed of the Pelton wheel, implies that the geometry is inherently a "low gear" design. Thus it is most suitable to being fed by a hydro source with a low ratio of flow to pressure, (meaning relatively low flow and/or relatively high pressure).

The specific speed is the main criterion for matching a specific hydro-electric site with the optimal turbine type. It also allows a new turbine design to be scaled from an existing design of known performance.

$$\eta_s = n\sqrt{P} / \sqrt{\rho}(gH)^{5/4} \text{ (dimensioned parameter)},$$

where:

- n = Frequency of rotation (rpm)
- P = Power (W)
- H = Water head (m)
- ρ = Density (kg/m³)

The formula implies that the Pelton turbine is *geared* most suitably for applications with relatively high hydraulic head H, due to the 5/4 exponent being greater than unity, and given the characteristically low specific speed of the Pelton.

Turbine Physics and Derivation

Energy and Initial Jet Velocity

In the ideal (frictionless) case, all of the hydraulic potential energy ($E_p = mgh$) is converted into kinetic energy ($E_k = mv^2/2$). Equating these two equations and solving for the initial jet velocity (V_i) indicates that the theoretical (maximum) jet velocity is $V_i = \sqrt{(2gh)}$. For simplicity, assume that all of the velocity vectors are parallel to each other. Defining the velocity of the wheel runner as: (u), then as the jet approaches the runner, the initial jet velocity relative to the runner is: ($V_i - u$). The initial jet velocity of jet is V_i

Final Jet Velocity

Assuming that the jet velocity is higher than the runner velocity, if the water is not to become backed-up in runner, then due to conservation of mass, the mass entering the runner must equal the mass leaving the runner. The fluid is assumed to be incompressible (an accurate assumption for most liquids). Also it is assumed that the cross-sectional area of the jet is constant. The jet *speed* remains constant relative to the runner. So as the jet recedes from the runner, the jet velocity relative to the runner is: $-(V_i - u) = -V_i + u$. In the standard reference frame (relative to the earth), the final velocity is then: $V_f = (-V_i + u) + u = -V_i + 2u$.

Optimal Wheel Speed

We know that the ideal runner speed will cause all of the kinetic energy in the jet to be transferred to the wheel. In this case the final jet velocity must be zero. If we let $-V_i + 2u = 0$, then the optimal runner speed will be $u = V_i/2$, or half the initial jet velocity.

Torque

By Newton's second and third laws, the force F imposed by the jet on the runner is equal but opposite to the rate of momentum change of the fluid, so:

$$F = -m(V_f - V_i) = -\rho Q[(-V_i + 2u) - V_i] = -\rho Q(-2V_i + 2u) = 2\rho Q(V_i - u)$$

where (ρ) is the density and (Q) is the volume rate of flow of fluid. If (D) is the wheel diameter, the torque on the runner is: $T = F(D/2) = \rho QD(V_i - u)$. The torque is at a maximum when the runner is stopped (i.e. when $u = 0$, $T = \rho QDV_i$). When the speed of the runner is equal to the initial jet velocity, the torque is zero (i.e. when $u = V_i$, then $T = 0$). On a plot of torque versus runner speed, the torque curve is straight between these two points, $(0, \rho QDV_i)$ and $(V_i, 0)$.

Power

The power $P = Fu = T\omega$, where ω is the angular velocity of the wheel. Substituting for F, we have $P = 2\rho Q(V_i - u)u$. To find the runner speed at maximum power, take the derivative of P with respect to u and set it equal to zero, $[dP/du = 2\rho Q(V_i - 2u)]$. Maximum power occurs when $u = V_i /2$. $P_{max} = \rho QV_i^2/2$. Substituting the initial jet power $V_i = \sqrt{(2gh)}$, this simplifies to $P_{max} = \rho ghQ$. This quantity exactly equals the kinetic power of the jet, so in this ideal case, the efficiency is 100%, since all the energy in the jet is converted to shaft output.

Efficiency

A wheel power divided by the initial jet power, is the turbine efficiency, $\eta = 4u(V_i - u)/V_i^2$. It is zero for $u = 0$ and for $u = V_i$. As the equations indicate, when a real Pelton wheel is working close to maximum efficiency, the fluid flows off the wheel with very little residual velocity. In theory, the energy efficiency varies only with the efficiency of the nozzle and wheel, and does not vary with hydraulic head. The term "efficiency" can refer to: Hydraulic, Mechanical, Volumetric, Wheel, or overall efficiency.

System Components

The conduit bringing high-pressure water to the impulse wheel is called the penstock. Originally the penstock was the name of the valve, but the term has been extended to include all of the fluid supply hydraulics. Penstock is now used as a general term for a water passage and control that is under pressure, whether it supplies an impulse turbine or not.

Turgo Turbine

The Turgo turbine is an impulse water turbine designed for medium head applications. Operational Turgo Turbines achieve efficiencies of about 87%. In factory and lab tests Turgo Turbines perform with efficiencies of up to 90%. It works with net heads between 15 and 300 m.

Developed in 1919 by Gilkes as a modification of the Pelton wheel, the Turgo has some advantages over Francis and Pelton designs for certain applications.

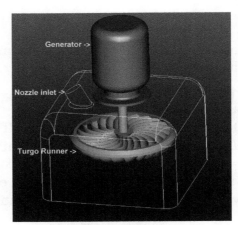

Turgo turbine and generator

First, the runner is less expensive to make than a Pelton wheel. Second, it doesn't need an airtight housing like the Francis. Third, it has higher specific speed and can handle a greater flow than the same diameter Pelton wheel, leading to reduced generator and installation cost.

Turgos operate in a head range where the Francis and Pelton overlap. While many large Turgo installations exist, they are also popular for small hydro where low cost is very important.

Like all turbines with nozzles, blockage by debris must be prevented for effective operation.

At Milford sound, New Zealand

Theory of Operation

The Turgo turbine is an impulse type turbine; water does not change pressure as it moves through the turbine blades. The water's potential energy is converted to kinetic energy with a nozzle. The high speed water jet is then directed on the turbine blades which deflect and reverse the flow. The resulting impulse spins the turbine runner, imparting energy to the turbine shaft. Water exits with very little energy. Turgo runners are extremely efficient

A Turgo runner looks like a Pelton runner split in half. For the same power, the Turgo runner is one half the diameter of the Pelton runner, and so twice the specific speed. The Turgo can handle a greater water flow than the Pelton because exiting water doesn't interfere with adjacent buckets.

The specific speed of Turgo runners is between the Francis and Pelton. Single or multiple nozzles can be used. Increasing the number of jets increases the specific speed of the runner by the square root of the number of jets (four jets yield twice the specific speed of one jet on the same turbine).

References

- Burns, Robert I. (1996), "Paper comes to the West, 800–1400", in Lindgren, Uta, Europäische Technik im Mittelalter. 800 bis 1400. Tradition und Innovation (4th ed.), Berlin: Gebr. Mann Verlag, pp. 413–422, ISBN 3-7861-1748-9

- de Crespigny, Rafe (2007), A Biographical Dictionary of Later Han to the Three Kingdoms (23-220 AD), Leiden: Koninklijke Brill, ISBN 90-04-15605-4

- Gimpel, Jean (1977), The Medieval Machine: The Industrial Revolution of the Middle Ages, London: Penguin (Non-Classics), ISBN 978-0-14-004514-7

- Langdon, John (2004), Mills in the Medieval Economy: England, 1300-1540, Oxford: Oxford University Press, ISBN 0-19-926558-5

- McErlean, Thomas; Crothers, Norman (2007), Harnessing the Tides: The Early Medieval Tide Mills at Nendrum Monastery, Strangford Lough, Belfast: Stationery Office Books, ISBN 978-0-337-08877-3

- Needham, Joseph. (1986). Science and Civilisation in China: Volume 4, Physics and Physical Technology; Part 2, Mechanical Engineering. Taipei: Caves Books Ltd. ISBN 0-521-05803-1.

- Pacey, Arnold, Technology in World Civilization: A Thousand-year History, The MIT Press; Reprint edition (July 1, 1991). ISBN 0-262-66072-5.

- Reynolds, Terry S. Stronger Than a Hundred Men: A History of the Vertical Water Wheel. (Johns Hopkins University Press 1983). ISBN 0-8018-7248-0.

- Rynne, Colin (2000), "Waterpower in Medieval Ireland", in Squatriti, Paolo, Working with Water in Medieval Europe, Technology and Change in History, 3, Leiden: Brill, pp. 1–50, ISBN 90-04-10680-4

- Wikander, Örjan (2000), "The Water-Mill", in Wikander, Örjan, Handbook of Ancient Water Technology, Technology and Change in History, 2, Leiden: Brill, pp. 371–400, ISBN 90-04-11123-9

- Thomson, Ross (2009). Structures of Change in the Mechanical Age: Technological Invention in the United States 1790-1865. Baltimore, MD: The Johns Hopkins University Press. p. 34. ISBN 978-0-8018-9141-0.

- Örjan Wikander (2008). "Chapter 6: Sources of Energy and Exploitation of Power". In John Peter Oleson. The Oxford Handbook of Engineering and Technology in the Classical World. Oxford University Press. pp. 141–2. ISBN 0-19-518731-8.

- Adriana de Miranda (2007). Water architecture in the lands of Syria: the water-wheels. L'Erma di Bretschneider. pp. 38–9. ISBN 88-8265-433-8.

- Donald Routledge Hill (1996), A history of engineering in classical and medieval times, Routledge, pp. 145–6, ISBN 0-415-15291-7

- *Nevell, Mike; Walker (2001). Portland Basin and the archaeology of the Canal Warehouse. Tameside Metropolitan Borough with University of Manchester Archaeological Unit. ISBN 1-871324-25-4.

- C Rossi; F Russo; F Russo (2009). "Ancient Engineers' Inventions: Precursors of the Present". Springer. ISBN 904812252X.

- Société d'énergie de la Baie James (1996). Le complexe hydroélectrique de La Grande Rivière : Réalisation de la deuxième phase (in French). Montreal: Société d'énergie de la Baie James. p. 397. ISBN 2-921077-27-2.

An Integrated Study of Marine Energy

Marine energy is the energy that is carried by ocean waves and ocean temperature differences. The energy created is then used to power homes and industries. The topics discussed in this section are marine current power, osmotic power, ocean thermal energy conversion and wave power. The topics discussed in the chapter are of great importance to broaden the existing knowledge on marine energy.

Marine Energy

Marine energy or marine power (also sometimes referred to as ocean energy, ocean power, or marine and hydrokinetic energy) refers to the energy carried by ocean waves, tides, salinity, and ocean temperature differences. The movement of water in the world's oceans creates a vast store of kinetic energy, or energy in motion. This energy can be harnessed to generate electricity to power homes, transport and industries.

The term marine energy encompasses both wave power i.e. power from surface waves, and tidal power i.e. obtained from the kinetic energy of large bodies of moving water. Offshore wind power is not a form of marine energy, as wind power is derived from the wind, even if the wind turbines are placed over water.

The oceans have a tremendous amount of energy and are close to many if not most concentrated populations. Ocean energy has the potential of providing a substantial amount of new renewable energy around the world.Energy from the ocean is also known as hydroelectricity.

Global Potential

There is the potential to develop 20,000–80,000 terawatt-hours (TWh) of electricity generated by changes in ocean temperatures, salt content, movements of tides, currents, waves and swells

Global potential	
Form	Annual generation
Tidal energy	>300 TWh
Marine current power	>800 TWh
Osmotic power Salinity gradient	2,000 TWh
Ocean thermal energy Thermal gradient	10,000 TWh
Wave energy	8,000–80,000 TWh
Source: IEA-OES, Annual Report 2007	

Indonesia as archipelagic country with three quarter of the area is ocean, has 49 GW recognized potential ocean energy and has 727 GW theoretical potential ocean energy.

Forms of Ocean Energy

Renewable

The oceans represent a vast and largely untapped source of energy in the form of surface waves, fluid flow, salinity gradients, and thermal.

Marine and Hydrokinetic (MHK) or marine energy development in U.S. and international waters includes projects using the following devices:

- Wave power converters in open coastal areas with significant waves;

- Tidal turbines placed in coastal and estuarine areas;

- In-stream turbines in fast-moving rivers;

- Ocean current turbines in areas of strong marine currents;

- Ocean thermal energy converters in deep tropical waters.

Marine Current Power

Strong ocean currents are generated from a combination of temperature, wind, salinity, bathymetry, and the rotation of the Earth. The Sun acts as the primary driving force, causing winds and temperature differences. Because there are only small fluctuations in current speed and stream location with no changes in direction, ocean currents may be suitable locations for deploying energy extraction devices such as turbines.

Ocean currents are instrumental in determining the climate in many regions around the world. While little is known about the effects of removing ocean current energy, the impacts of removing current energy on the farfield environment may be a significant environmental concern. The typical turbine issues with blade strike, entanglement of marine organisms, and acoustic effects still exists; however, these may be magnified due to the presence of more diverse populations of marine organisms using ocean currents for migration purposes. Locations can be further offshore and therefore require longer power cables that could affect the marine environment with electromagnetic output.

Osmotic Power

At the mouth of rivers where fresh water mixes with salt water, energy associated with the salinity gradient can be harnessed using pressure-retarded reverse osmosis process and associated conversion technologies. Another system is based on using freshwater upwelling through a turbine immersed in seawater, and one involving electrochemical reactions is also in development.

Significant research took place from 1975 to 1985 and gave various results regarding the economy of PRO and RED plants. It is important to note that small-scale investigations into salinity power production take place in other countries like Japan, Israel, and the United States. In Europe the research is concentrated in Norway and the Netherlands, in both places small pilots are tested. Salinity gradient energy is the energy available from the difference in salt concentration between freshwater with saltwater. This energy source is not easy to understand, as it is not directly occurring in nature in the form of heat, waterfalls, wind, waves, or radiation.

Ocean Thermal Energy

Water typically varies in temperature from the surface warmed by direct sunlight to greater depths where sunlight cannot penetrate. This differential is greatest in tropical waters, making this technology most applicable in water locations. A fluid is often vaporized to drive a turbine that may generate electricity or produce desalinized water. Systems may be either open-cycle, closed-cycle, or hybrid.

Tidal Power

The energy from moving masses of water — a popular form of hydroelectric power generation. Tidal power generation comprises three main forms, namely: tidal stream power, tidal barrage power, and dynamic tidal power.

Wave Power

Solar energy from the Sun creates temperature differentials that result in wind. The interaction between wind and the surface of water creates waves, which are larger when there is a greater distance for them to build up. Wave energy potential is greatest between 30° and 60° latitude in both hemispheres on the west coast because of the global direction of wind. When evaluating wave energy as a technology type, it is important to distinguish between the four most common approaches: point absorber buoys, surface attenuators, oscillating water columns, and overtopping devices.

The wave energy sector is reaching a significant milestone in the development of the industry, with positive steps towards commercial viability being taken. The more advanced device developers are now progressing beyond single unit demonstration devices and are proceeding to array development and multi-megawatt projects. The backing of major utility companies is now manifesting itself through partnerships within the development process, unlocking further investment and, in some cases, international co-operation.

At a simplified level, wave energy technology can be located near-shore and offshore. Wave energy converters can also be designed for operation in specific water depth conditions: deep water, intermediate water or shallow water. The fundamental device design will be dependent on the location of the device and the intended resource characteristics.

Non-renewable

Petroleum and natural gas beneath the ocean floor are also sometimes considered a form of ocean energy. An ocean engineer directs all phases of discovering, extracting, and delivering offshore petroleum (via oil tankers and pipelines,) a complex and demanding task. Also centrally important is the development of new methods to protect marine wildlife and coastal regions against the undesirable side effects of offshore oil extraction.

Marine Energy Development

The UK is leading the way in wave and tidal (marine) power generation. The world's first marine energy test facility was established in 2003 to kick start the development of the marine energy industry in the UK. Based in Orkney, Scotland, the European Marine Energy Centre (EMEC) has supported the deployment of more wave and tidal energy devices than at any other single site in the world. The Centre was established with around £36 million of funding from the Scottish Government, Highlands and Islands Enterprise, the Carbon Trust, UK Government, Scottish Enterprise, the European Union and Orkney Islands Council, and is the only accredited wave and tidal test centre for marine renewable energy in the world, suitable for testing a number of full-scale devices simultaneously in some of the harshest weather conditions while producing electricity to the national grid.

Clients that have tested at the centre include Aquamarine Power, AW Energy, Pelamis Wave Power, Seatricity, ScottishPower Renewables and Wello on the wave site, and Alstom (formerly Tidal Generation Ltd), ANDRITZ HYDRO Hammerfest, Kawasaki Heavy Industries, Magallanes, Nautricity, Open Hydro, Scotrenewables Tidal Power, and Voith on the tidal site.

Leading the €11m FORESEA (Funding Ocean Renewable Energy through Strategic European Action) project, which provides funding support to ocean energy technology developers to access Europe's world-leading ocean energy test facilities, EMEC will welcome a number of wave and tidal clients to their pipeline for testing on site.

Beyond device testing, EMEC also provides a wide range of consultancy and research services, and is working closely with Marine Scotland to streamline the consenting process for marine energy developers. EMEC is at the forefront in the development of international standards for marine energy, and is forging alliances with other countries, exporting its knowledge around the world to stimulate the development of a global marine renewables industry.

Environmental Effects

Common environmental concerns associated with marine energy developments include:

- the risk of marine mammals and fish being struck by tidal turbine blades
- the effects of EMF and underwater noise emitted from operating marine energy devices
- the physical presence of marine energy projects and their potential to alter the behavior of marine mammals, fish, and seabirds with attraction or avoidance
- the potential effect on nearfield and farfield marine environment and processes such as sediment transport and water quality

The Tethys database provides access to scientific literature and general information on the potential environmental effects of marine energy.

Marine Current Power

Marine current power is a form of marine energy obtained from harnessing of the kinetic energy of marine currents, such as the Gulf stream. Although not widely used at present, marine current power has an important potential for future electricity generation. Marine currents are more predictable than wind and solar power.

A 2006 report from United States Department of the Interior estimates that capturing just $1/_{1,000th}$ of the available energy from the Gulf Stream would supply Florida with 35% of its electrical needs.

Strong ocean currents are generated from a combination of temperature, wind, salinity, bathymetry, and the rotation of the earth. The sun acts as the primary driving force, causing winds and temperature differences. Because there are only small fluctuations in current speed and stream location with minimal changes in direction, ocean currents may be suitable locations for deploying energy extraction devices such as turbines. Other effects such as regional differences in temperature and salinity and the Coriolis effect due to the rotation of the earth are also major influences. The kinetic energy of marine currents can be converted in much the same way that a wind turbine extracts energy from the wind, using various types of open-flow rotors.

The potential of electric power generation from marine tidal currents is enormous. There are several factors that make electricity generation from marine currents very appealing when compared to other renewables:

- The high load factors resulting from the fluid properties. The predictability of the resource, so that, unlike most of other renewables, the future availability of energy can be known and planned for.

- The potentially large resource that can be exploited with little environmental impact, thereby offering one of the least damaging methods for large-scale electricity generation.

- The feasibility of marine-current power installations to provide also base grid power, especially if two or more separate arrays with offset peak-flow periods are interconnected.

Early Experiences

The possible use of marine currents as an energy resource began to draw attention in the mid-1970s after the first oil crisis. In 1974 several conceptual designs were presented at the MacArthur Workshop on Energy, and in 1976 the British General Electric Co. undertook a partially government-funded study which concluded that marine current power deserved more detailed research. Soon after, the ITD-Group in UK implemented a research program involving a year performance testing of a 3-m hydroDarrieus rotor deployed at Juba on the White Nile.

The 1980s saw a number of small research projects to evaluate marine current power systems. The main countries where studies were carried out were the UK, Canada, and Japan. In 1992–1993 the Tidal Stream Energy Review identified specific sites in UK waters with suitable current speed to generate up to 58 TWh/year. It confirmed a total marine current power resource capable theoretically of meeting some 19% of the UK electricity demand.

In 1994–1995 the EU-JOULE CENEX project identified over 100 European sites ranging from 2 to 200 km² of sea-bed area, many with power densities above 10 MW/km². Both the UK Government and the EU have committed themselves to internationally negotiated agreements designed to combat global warming. In order to comply with such agreements, an increase in large-scale electricity generation from renewable resources will be required. Marine currents have the potential to supply a substantial share of future EU electricity needs. The study of 106 possible sites for tidal turbines in the EU showed a total potential for power generation of about 50 TWh/year. If this resource is to be successfully utilized, the technology required could form the basis of a major new industry to produce clean power for the 21st century.

Alternative Technologies in Marine-current-power Applications

There are several types of open-flow devices that can be used in marine-current-power applications; many of them are modern descendants of the waterwheel or similar. However, the more technically sophisticated designs, derived from wind-power rotors, are the most likely to achieve enough cost-effectiveness and reliability to be practical in a massive marine-current-power future scenario. Even though there is no generally accepted term for these open-flow hydro-turbines, some sources refer to them as water-current turbines. There are two main types of water current turbines that might be considered: axial-flow horizontal-axis propellers (with both variable-pitch or fixed-pitch), and cross-flow vertical-axis Darrieus rotors. Both rotor types may be combined with any of the three main methods for supporting water-current turbines: floating moored systems, sea-bed mounted systems, and intermediate systems. Sea-bed-mounted monopile structures constitute the first-generation marine current power systems. They have the advantage of using existing (and reliable) engineering know-how, but they are limited to relatively shallow waters (about 20 to 40 m depth).

Environmental Effects

Ocean currents are instrumental in determining the climate in many regions around the world. While little is known about the effects of removing ocean current energy, the impacts of removing current energy on the farfield environment may be a significant environmental concern. The typical turbine issues with blade strike, entanglement of marine organisms, and acoustic effects still exists; however, these may be magnified due to the presence of more diverse populations of marine organisms using ocean currents for migration purposes. Locations can be further offshore and therefore require longer power cables that could affect the marine environment with electro-magnetic output.

The Tethys database provides access to scientific literature and general information on the potential environmental effects of ocean current energy.

Osmotic Power

Osmotic power, salinity gradient power or blue energy is the energy available from the difference in the salt concentration between seawater and river water. Two practical methods for this are reverse electrodialysis (RED) and pressure retarded osmosis (PRO). Both processes rely on osmosis with ion specific membranes. The key waste product is brackish water. This byproduct is the result of natural forces that are being harnessed: the flow of fresh water into seas that are made up of salt water.

In 1954 Pattle suggested that there was an untapped source of power when a river mixes with the sea, in terms of the lost osmotic pressure, however it was not until the mid '70s where a practical method of exploiting it using selectively permeable membranes by Loeb was outlined.

The method of generating power by pressure retarded osmosis was invented by Prof. Sidney Loeb in 1973 at the Ben-Gurion University of the Negev, Beersheba, Israel. The idea came to Prof. Loeb, in part, as he observed the Jordan River flowing into the Dead Sea. He wanted to harvest the energy of mixing of the two aqueous solutions (the Jordan River being one and the Dead Sea being the other) that was going to waste in this natural mixing process. In 1977 Prof. Loeb invented a method of producing power by a reverse electrodialysis heat engine.

The technologies have been confirmed in laboratory conditions. They are being developed into commercial use in the Netherlands (RED) and Norway (PRO). The cost of the membrane has been an obstacle. A new, lower cost membrane, based on an electrically modified polyethylene plastic, made it fit for potential commercial use. Other methods have been proposed and are currently under development. Among them, a method based on electric double-layer capacitor technology. and a method based on vapor pressure difference.

Basics of Salinity Gradient Power

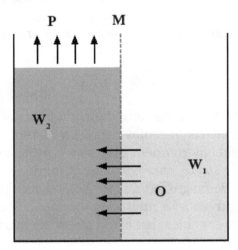

Pressure-retarded osmosis

Salinity gradient power is a specific renewable energy alternative that creates renewable and sustainable power by using naturally occurring processes. This practice does not contaminate or release carbon dioxide (CO_2) emissions (vapor pressure methods will release dissolved air containing CO_2 at low pressures—these non-condensable gases can be re-dissolved of course, but with an energy penalty). Also as stated by Jones and Finley within their article "Recent Development in Salinity Gradient Power", there is basically no fuel cost.

Salinity gradient energy is based on using the resources of "osmotic pressure difference between fresh water and sea water." All energy that is proposed to use salinity gradient technology relies on the evaporation to separate water from salt. Osmotic pressure is the "chemical potential of concentrated and dilute solutions of salt". When looking at relations between high osmotic pressure and low, solutions with higher concentrations of salt have higher pressure.

Differing salinity gradient power generations exist but one of the most commonly discussed is pressure-retarded osmosis (PRO). Within PRO seawater is pumped into a pressure chamber where the pressure is lower than the difference between fresh and salt water pressure. Fresh water moves in a semipermeable membrane and increases its volume in the chamber. As the pressure in the chamber is compensated a turbine spins to generate electricity. In Braun's article he states that this process is easy to understand in a more broken down manner. Two solutions, A being salt water and B being fresh water are separated by a membrane. He states "only water molecules can pass the semipermeable membrane. As a result of the osmotic pressure difference between both solutions, the water from solution B thus will diffuse through the membrane in order to dilute solution A". The pressure drives the turbines and power the generator that produces the electrical energy. Osmosis might be used directly to "pump" fresh water out of The Netherlands into the sea. This is currently done using electric pumps.

Efficiency

A 2012 study on efficiency from Yale university concluded that the highest extractable work in constant-pressure PRO with a seawater draw solution and river water feed solution is 0.75 kWh/m³ while the free energy of mixing is 0.81 kWh/m³—a thermodynamic extraction efficiency of 91.0%.

Methods

While the mechanics and concepts of salinity gradient power are still being studied, the power source has been implemented in several different locations. Most of these are experimental, but thus far they have been predominantly successful. The various companies that have utilized this power have also done so in many different ways as there are several concepts and processes that harness the power from salinity gradient.

Pressure-Retarded Osmosis

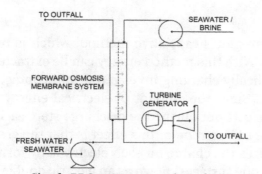

Simple PRO power generation scheme

One method to utilize salinity gradient energy is called pressure-retarded osmosis. In this method, seawater is pumped into a pressure chamber that is at a pressure lower than the difference between the pressures of saline water and fresh water. Freshwater is also pumped into the pressure chamber through a membrane, which increase both the volume and pressure of the chamber. As the pressure differences are compensated, a turbine is spun, providing kinetic energy. This method is being specifically studied by the Norwegian utility Statkraft, which has calculated that up to 2.85 GW would be available from this process in Norway. Statkraft has built the world's first prototype

PRO power plant on the Oslo fjord which was opened by Princess Mette-Marit of Norway on November 24, 2009. It aims to produce enough electricity to light and heat a small town within five years by osmosis. At first it will produce a minuscule 4 kilowatts – enough to heat a large electric kettle, but by 2015 the target is 25 megawatts – the same as a small wind farm.

Osmotic Power Prototype at Tofte (Hurum), Norway

Reversed Electrodialysis

A second method being developed and studied is reversed electrodialysis or reverse dialysis, which is essentially the creation of a salt battery. This method was described by Weinstein and Leitz as "an array of alternating anion and cation exchange membranes can be used to generate electric power from the free energy of river and sea water."

The technology related to this type of power is still in its infant stages, even though the principle was discovered in the 1950s. Standards and a complete understanding of all the ways salinity gradients can be utilized are important goals to strive for in order make this clean energy source more viable in the future.

Capacitive Method

A third method is Doriano Brogioli's capacitive method, which is relatively new and has so far only been tested on lab scale. With this method energy can be extracted out of the mixing of saline water and freshwater by cyclically charging up electrodes in contact with saline water, followed by a discharge in freshwater. Since the amount of electrical energy which is needed during the charging step is less than one gets out during the discharge step, each completed cycle effectively produces energy. An intuitive explanation of this effect is that the great number of ions in the saline water efficiently neutralizes the charge on each electrode by forming a thin layer of opposite charge very close to the electrode surface, known as an electric double layer. Therefore, the voltage over the electrodes remains low during the charge step and charging is relatively easy. In between the charge and discharge step, the electrodes are brought in contact with freshwater. After this, there are less ions available to neutralize the charge on each electrode such that the voltage over the electrodes increases. The discharge step which follows is therefore able to deliver a relatively high amount of energy. A physical explanation is that on an electrically charged capacitor, there is a mutually attractive electric force between the electric charge on the electrode, and the ionic charge in the liquid. In order to pull ions away from the charged electrode, osmotic pressure must do work. This work done increases the electrical potential energy in the capacitor. An electronic explanation is that capacitance is a function of ion density. By introducing a salinity gradient and allowing some of the ions to diffuse out of the capacitor, this reduces the capacitance, and so the voltage must increase, since the voltage equals the ratio of charge to capacitance.

Vapor Pressure Differences: Open Cycle and Absorption Refrigeration Cycle (Closed Cycle)

Both of these methods do not rely on membranes, so filtration requirements are not as important as they are in the PRO & RED schemes.

Open Cycle

Similar to the open cycle in ocean thermal energy conversion (OTEC). The disadvantage of this cycle is the cumbersome problem of a large diameter turbine (75 meters +) operating at below atmospheric pressure to extract the power between the water with less salinity & the water with greater salinity.

Absorption Refrigeration Cycle (Closed Cycle)

For the purpose of dehumidifying air, in a water-spray absorption refrigeration system, water vapor is dissolved into a deliquescent salt water mixture using osmotic power as an intermediary. The primary power source originates from a thermal difference, as part of a thermodynamic heat engine cycle.

Solar Pond

At the Eddy Potash Mine in New Mexico, a technology called "salinity gradient solar pond" (SGSP) is being utilized to provide the energy needed by the mine. This method does not harness osmotic power, only solar power. Sunlight reaching the bottom of the saltwater pond is absorbed as heat. The effect of natural convection, wherein "heat rises", is blocked using density differences between the three layers that make up the pond, in order to trap heat. The upper convection zone is the uppermost zone, followed by the stable gradient zone, then the bottom thermal zone. The stable gradient zone is the most important. The saltwater in this layer can not rise to the higher zone because the saltwater above has lower salinity and is therefore less-dense and more buoyant; and it can not sink to the lower level because that saltwater is denser. This middle zone, the stable gradient zone, effectively becomes an "insulator" for the bottom layer (although the main purpose is to block natural convection, since water is a poor insulator). This water from the lower layer, the storage zone, is pumped out and the heat is used to produce energy, usually by turbine in an organic Rankine cycle.

In theory a solar pond *could* be used to generate osmotic power if evaporation from solar heat is used to create a salinity gradient, *and* the potential energy in this salinity gradient is *harnessed directly* using one of the first three methods above, such as the capacitive method.

Boron Nitride Nanotubes

A research team built an experimental system using boron nitride that produced much greater power than the Statoil prototype. It used an impermeable and electrically insulating membrane that was pierced by a single boron nitride nanotube with an external diameter of a few dozen nanometers. With this membrane separating a salt water reservoir and a fresh water reservoir, the team measured the electric current passing through the membrane using two electrodes immersed in the fluid either side of the nanotube.

The results showed the device was able to generate an electric current on the order of a nanoampere. The researchers claim this is 1,000 times the yield of other known techniques for harvesting osmotic energy and makes boron nitride nanotubes an extremely efficient solution for harvesting the energy of salinity gradients for usable electrical power.

The team claimed that a 1 square metre (11 sq ft) membrane could generate around 4 kW and be capable of generating up to 30 MWh per year.

Possible Negative Environmental Impact

Marine and river environments have obvious differences in water quality, namely salinity. Each species of aquatic plant and animal is adapted to survive in either marine, brackish, or freshwater environments. There are species that can tolerate both, but these species usually thrive best in a specific water environment. The main waste product of salinity gradient technology is brackish water. The discharge of brackish water into the surrounding waters, if done in large quantities and with any regularity, will cause salinity fluctuations. While some variation in salinity is usual, particularly where fresh water (rivers) empties into an ocean or sea anyway, these variations become less important for both bodies of water with the addition of brackish waste waters. Extreme salinity changes in an aquatic environment may result in findings of low densities of both animals and plants due to intolerance of sudden severe salinity drops or spikes. According to the prevailing environmentalist opinions, the possibility of these negative effects should be considered by the operators of future large blue energy establishments.

The impact of brackish water on ecosystems can be minimized by pumping it out to sea and releasing it into the mid-layer, away from the surface and bottom ecosystems.

Impingement and entrainment at intake structures are a concern due to large volumes of both river and sea water utilized in both PRO and RED schemes. Intake construction permits must meet strict environmental regulations and desalination plants and power plants that utilize surface water are sometimes involved with various local, state and federal agencies to obtain permission that can take upwards to 18 months.

Ocean Thermal Energy Conversion

Ocean Thermal Energy Conversion (OTEC) is a clean, zero-emission and renewable energy technology. OTEC takes the heat from tropical oceans and converts it to electricity. OTEC is capable of generating electricity day and night, throughout the year, providing a reliable source of electricity. Although still largely untapped, OTEC is one of the world's largest renewable energy resources and is available to around 100 countries within their nautical economical zone.

Potential

The ocean comprises an enormous energy source. Covering almost two-third of the surface of the earth, the ocean captures 70% of the solar energy that irradiates on earth. It is estimated that the solar energy that is absorbed by the ocean per year, exceeds the human energy consumption more than 4000 times. Recent research concludes that there is between 7 and 30 terawatt of electric potential energy available without having an adverse impact on natural thermal currents and ocean temperatures, this equals three to ten times the global electricity demand. This vast resource has been recognized worldwide in recent reports from the International Institute for Applied Systems Analysis (IIASA) and the International Panel on Climate Change.

An advantage of OTEC compared to other renewable energy sources is the reliable and predictable energy production. Since tropical oceans hardly encounter fluctuations in their surface temperature (neither per day, nor per season), the temperature difference between the various oceanic layers remains nearly constant. This enables OTEC to provide a base-load electricity supply with a capacity factor of 80% - 100%.

Oceanic Layered Structure

The oceans cover almost two-third of the surface of the earth and capture a majority of the solar energy that reaches the earth. Especially in tropical regions, solar energy is absorbed by the ocean and stored as heat. The balance between the incoming solar radiation and the heat loss due to evaporation and convection accounts for a constant temperature of the oceanic surface water.

As depth increases, the ocean water becomes colder, due to the accumulation of ice-cold water that has melted from the polar regions. Because of its higher density, this cold water flows along the bottom of the ocean from the poles towards the equator, displacing the lower-density water above. These two phenomena provide for a layered oceanic structure in deep, tropical oceans with a reservoir of warm water at the surface and a reservoir of cold water deeper in the ocean. The temperature difference between these layers varies between 22 °C and 25 °C. Until today this vast sustainable energy reservoir remains largely unused.

Cycle

Working Principle

The OTEC system is based on an organic Rankine cycle; a working fluid with a lower boiling point and a higher vapour pressure than water is used to power a turbine that generates electricity. First, warm water from the ocean surface is pumped through a heat exchanger. In the heat exchanger, the heat that is exchanged from the seawater to the working fluid causes the working fluid to vaporize. This vaporized working fluid is expanded in a turbine that is connected to a generator that generates electricity. Thereafter, cold seawater, pumped through a second heat exchanger, condenses the vapour into a liquid, so it can be reused. An electricity-generating cycle is therefore created.

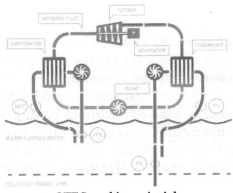

OTEC working principle

Working Fluids

Effective electricity generation with OTEC requires a working fluid with a lower boiling point and a higher vapour pressure than water. A typical choice of working fluid is ammonia, which has superior transport properties and is easily available at low cost. Also, the extensive operational experience with ammonia in refrigeration systems and its proven safety record make it the preferred choice of various working fluids, such as propane and other refrigerants. The working fluid is contained in a closed system, at relative low operating pressures and temperatures. Much lower than in for instance fossil fuel or nuclear power plants. Nonetheless, sealing of the components that contain the working fluid needs to be taken care of, but reliable solutions are readily available.

Efficiency

In line with the Carnot efficiency, a heat engine gives greater efficiency when run with a large temperature difference. The temperature difference between the surface and deep water of the ocean is greatest in the tropics, although still a modest 20 to 25 °C. It is therefore in the tropics that OTEC offers the greatest possibilities. The energy consumption of an OTEC cycle is dominated by the seawater pumps. These pumps and other auxiliary equipment consume roughly 20% of the total electricity produced. The remaining 80% is net power and can then be supplied to the grid.

Cost and Economics

Because OTEC systems have not been widely deployed yet, cost estimates are uncertain. The current Levelized Cost of Energy (LCOE) of OTEC is estimated at about $0.20 -$0.25 per kWh, with significant room for cost reductions to even lower than $0.07 per kWh once the systems become more mature. The LCOE of OTEC is predominantly determined by the initial capital investment. The pipes to transport the deep seawater and the heat exchangers form one of the largest capital investments of OTEC systems. Annual operation and maintenance cost are about 1% of the capital investment.

Evironmental Impact

The energy produced by OTEC is clean, zero-emission and renewable. It will drastically reduce emissions and make energy available from an inexhaustible natural resource. A life cycle analysis of a OTEC plant, assuming currently available technology, resulted in a global warming potential of OTEC that is at most 3% of diesel generated electricity and its energy payback is within 1 to 2 years. It is anticipated that this figure will be further improved by improving technology.

OTEC requires seawater flow rates of several cubic meters per second per net megawatt of electricity produced. Though substantial, these flow rates are negligible compared to normal ocean currents with flow rates of many million cubic meters per second. By selecting the right location for the seawater intakes and the size of mesh for the intake filters, the possible entrainment of organisms is minimized. Generally speaking, the problem can often be reduced by placing the seawater intake further from the shore while avoiding submarine canyons, coral reefs or areas with fast ocean currents.

The seawater coming out of the OTEC plant is returned to a level in the ocean with approximately the same temperature and below the photic zone. The latter ensures that the discharge plume with nutrient-rich deep seawater doesn't trigger biological growth. The exact siting of the discharge pipe will vary according to currents and temperatures at the specific location. It is typically around several tens to two hundred meter deep.

Most recently, NOAA held an OTEC Workshop in 2010 and 2012, seeking to assess the physical, chemical, and biological impacts and risks of OTEC, and to identify information gaps or needs . Today's available environmental modeling tools, sensors and monitoring techniques greatly help in analyzing and monitoring impact at specific locations. The Tethys database provides access to scientific literature and general information on the potential environmental effects of OTEC.

OTEC Around the World

Locations

In order to be operational, OTEC requires a temperature difference of the seawater of 20 °C or more. OTEC resource exists therefore in tropical waters, typically in the equatorial region between 20°N and 20°S. Recent research identified 98 nations and territories with OTEC resources within their nautical economical zone.

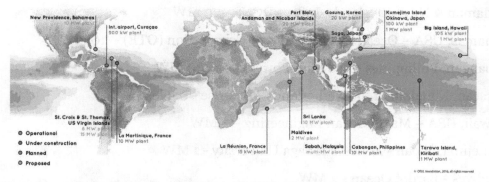

Schematic of the OTEC potential around the world (in colour) and the OTEC plants that are built or planned (source: OTEC Foundation)

OTEC Plants in Operation

OTEC projects around the world. 1. top left: Hawaii, US (105 kW), 2. mid left: Delft, Netherlands (lab-scale), 3. bottom left: La Réunion (15 kW), 4. top right: Kumejima, Japan (100 kW), 5. mid right: Saga, Japan (30 kW), 6. bottom right: Korea (20 kW). (source: OTEC Foundation)

- Saga, Japan - Xenesys & Saga University - 30 kW - operational since 1980 with the purpose of research and development

- Gosung, Korea - KRISO - 20 kW - operational since 2012 with the purpose of research and development

- Réunion Island, France - DCNS - 15 kW - operational since 2012 with the purpose of research and development

- Kumejima, Japan - Xenesys & Saga University - 100 KW - grid connected - operational since 2013 with the purpose of research and development and for electricity production

- Hawaii, US - Makai Ocean Engineering - 105 kW - grid connected - operational since 2015 with the purpose of electricity production

OTEC Plants Under Development

- Andaman & Nicobar Islands, India - DCNS - 20 MW

- Bahamas, USA – Ocean Thermal Energy Corporation (OTE) – 10 MW

- Cabangan, Philippines - Bell Pirie Power Corp - 10 MW

- Curaçao, Kingdom of the Netherlands - Bluerise - 0.5 MW

- Hawaii, USA – Makai Ocean Engineering – 1 MW

- Kumejima, Japan - Xenesys & Saga University - 1 MW

- Maldives - Bardot Ocean - 2 MW

- Martinique, France – Akuoa Energy & DCNS – 10,7 MW

- Sri Lanka - Bluerise - 10 MW

- Tarawa Island, Kiribati– [KRISO] – 1 MW

- US Virgin Islands - Ocean Thermal Energy Corporation (OTE) – 8 & 15 MW

History

The concept of generating electricity by means of the temperature difference between the various layers in the ocean was first mentioned by Jules Verne in 1870. In his book "Twenty Thousand Leagues Under the Sea", he used the concept to power the submarine of Captain Nemo. Attempts to develop and refine OTEC technology started in the 1880s. In 1881, Jacques Arsene d'Arsonval, a French physicist, proposed tapping the thermal energy of the ocean. D'Arsonval's student, Georges Claude, built the first OTEC plant, in Matanzas, Cuba in 1930 . The system generated 22 kW of electricity with a low-pressure turbine. The plant was later destroyed in a storm. In 1935, Claude constructed a plant aboard a 10,000-ton cargo vessel moored off the coast of Brazil. Weather and waves destroyed it before it could generate net power.

In the 50's and 60' several proposals were made for the installation of OTEC plants. Moreover, in 1962, J. Hilbert Anderson and James H. Anderson, Jr. focused on increasing component efficiency. They patented their new "closed cycle" design in 1967. This design improved upon the original closed-cycle Rankine system, and included this in an outline for a plant that would produce power at lower cost than oil or coal. However, little attention was drawn to OTEC, since coal and nuclear were considered the future of energy.

The 1970s saw an uptick in OTEC research, due to the oil crisis. The U.S. federal government invested $260 million in OTEC research after President Carter signed a law that committed the US to a production goal of 10,000 MW of electricity from OTEC systems by 1999. But a falling oil price smothered the interest in OTEC.

In the meantime, an OTEC plant was constructed on the island of Nauru in Japan. The installation became operational in 1981 en produced 120 kW of electricity; 90 kW to power the system and 30 kW for the island grid. 1981 also saw a major development in OTEC technology when Russian engineer, Dr. Alexander Kalina, instead of pure ammonia, used a mixture of ammonia and water as a working fluid to produce electricity. This new ammonia-water mixture greatly improved the efficiency of the power cycle.

With the start of the 21st century, interest in OTEC began to rise again. Currently, several OTEC systems up to 1 megawatt are operational and the first multi-megawatt installations are under development.

Related Technologies

The implementation of OTEC offers the possibility of co-generating other sustainable products. First of all,the warm and cold seawater can be used for efficient production of clean drinking water. Through evaporation of the warm seawater, the salt is removed. Subsequent condensation of the desalinated warm seawater then provides fresh drinking water. System analysis indicates that a 2-megawatt OTEC plant could produce about 4,300 cubic metres (150,000 cu ft) of desalinated water each day. Secondly, the cold seawater can be used in combination with Seawater District Cooling (SDC) for energy efficient air-conditioning of local buildings (also known as Sea Water Air Conditioning, or SWAC). SDC can provide around 90% savings of electricity when compared to traditional cooling systems. These are considerable savings, taking into account that an estimate of 40% of the energy consumed in developed tropical regions is cooling related. Furthermore, the cold seawater can be used in agriculture to cool soil or greenhouses and thereby enable the perfect climate for crops to grow faster and healthier. Moreover, the cold seawater from the deep ocean is rich in nutrients, such as phosphates and nitrates and is virtually free of pathogen. This makes it ideal for use in aquaculture. The cold water opens the possibilities of growing high value fishes that otherwise can not been grown locally. Similarly, technologies like algae, cosmetics and nutraceuticals production can benefit from the nutrients found in deep seawater.

Wave Power

Pelamis Wave Energy Converter on site at the European Marine Energy Centre (EMEC), in 2008

Azura at the US Navy's Wave Energy Test Site (WETS) on Oahu

SINN Power Wave Energy Converter

Wave power is the transport of energy by wind waves, and the capture of that energy to do useful work – for example, electricity generation, water desalination, or the pumping of water (into reservoirs). A machine able to exploit wave power is generally known as a wave energy converter (WEC).

Wave power is distinct from the diurnal flux of tidal power and the steady gyre of ocean currents. Wave-power generation is not currently a widely employed commercial technology, although there have been attempts to use it since at least 1890. In 2008, the first experimental wave farm was opened in Portugal, at the Aguçadoura Wave Park.

Physical Concepts

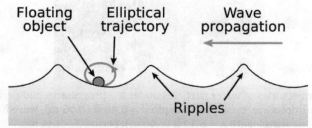

Floating object   Elliptical trajectory   Wave propagation

Ripples

When an object bobs up and down on a ripple in a pond, it follows approximately an elliptical trajectory.

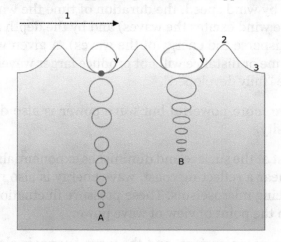

Motion of a particle in an ocean wave.
A = At deep water. The elliptical motion of fluid particles decreases rapidly with increasing depth below the surface.
B = At shallow water (ocean floor is now at B). The elliptical movement of a fluid particle flattens with decreasing depth.
1 = Propagation direction.
2 = Wave crest.
3 = Wave trough.

Waves are generated by wind passing over the surface of the sea. As long as the waves propagate slower than the wind speed just above the waves, there is an energy transfer from the wind to the waves. Both air pressure differences between the upwind and the lee side of a wave crest, as well as friction on the water surface by the wind, making the water to go into the shear stress causes the growth of the waves.

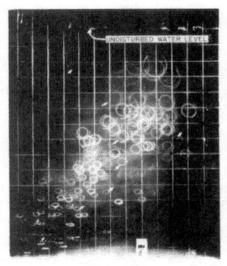

Photograph of the elliptical trajectories of water particles under a – progressive and periodic – surface gravity wave in a wave flume. The wave conditions are: mean water depth d = 2.50 ft (0.76 m), wave height H = 0.339 ft (0.103 m), wavelength λ = 6.42 ft (1.96 m), period T = 1.12 s.

Wave height is determined by wind speed, the duration of time the wind has been blowing, fetch (the distance over which the wind excites the waves) and by the depth and topography of the sea-floor (which can focus or disperse the energy of the waves). A given wind speed has a matching practical limit over which time or distance will not produce larger waves. When this limit has been reached the sea is said to be "fully developed".

In general, larger waves are more powerful but wave power is also determined by wave speed, wavelength, and water density.

Oscillatory motion is highest at the surface and diminishes exponentially with depth. However, for standing waves (clapotis) near a reflecting coast, wave energy is also present as pressure oscillations at great depth, producing microseisms. These pressure fluctuations at greater depth are too small to be interesting from the point of view of wave power.

The waves propagate on the ocean surface, and the wave energy is also transported horizontally with the group velocity. The mean transport rate of the wave energy through a vertical plane of unit width, parallel to a wave crest, is called the wave energy flux (or wave power, which must not be confused with the actual power generated by a wave power device).

Wave Power Formula

In deep water where the water depth is larger than half the wavelength, the wave energy flux is[a]

$$P = \frac{\rho g^2}{64\pi} H_{m0}^2 T_e \approx \left(0.5 \, \frac{\text{kW}}{\text{m}^3 \cdot \text{s}} \right) H_{m0}^2 \, T_e,$$

with P the wave energy flux per unit of wave-crest length, H_{mo} the significant wave height, T_e the wave energy period, ρ the water density and g the acceleration by gravity. The above formula states that wave power is proportional to the wave energy period and to the square of the wave height. When the significant wave height is given in metres, and the wave period in seconds, the result is the wave power in kilowatts (kW) per metre of wavefront length.

Example: Consider moderate ocean swells, in deep water, a few km off a coastline, with a wave height of 3 m and a wave energy period of 8 seconds. Using the formula to solve for power, we get

$$P \approx 0.5 \frac{\text{kW}}{\text{m}^3 \cdot \text{s}} (3 \cdot \text{m})^2 (8 \cdot \text{s}) \approx 36 \frac{\text{kW}}{\text{m}},$$

meaning there are 36 kilowatts of power potential per meter of wave crest.

In major storms, the largest waves offshore are about 15 meters high and have a period of about 15 seconds. According to the above formula, such waves carry about 1.7 MW of power across each metre of wavefront.

An effective wave power device captures as much as possible of the wave energy flux. As a result, the waves will be of lower height in the region behind the wave power device.

Wave Energy and Wave-Energy Flux

In a sea state, the average(mean) energy density per unit area of gravity waves on the water surface is proportional to the wave height squared, according to linear wave theory:

$$E = \frac{1}{16} \rho g H_{m0}^2,$$

where E is the mean wave energy density per unit horizontal area (J/m²), the sum of kinetic and potential energy density per unit horizontal area. The potential energy density is equal to the kinetic energy, both contributing half to the wave energy density E, as can be expected from the equipartition theorem. In ocean waves, surface tension effects are negligible for wavelengths above a few decimetres.

As the waves propagate, their energy is transported. The energy transport velocity is the group velocity. As a result, the wave energy flux, through a vertical plane of unit width perpendicular to the wave propagation direction, is equal to:

$$P = E c_g,$$

with c_g the group velocity (m/s). Due to the dispersion relation for water waves under the action of gravity, the group velocity depends on the wavelength λ, or equivalently, on the wave period T. Further, the dispersion relation is a function of the water depth h. As a result, the group velocity behaves differently in the limits of deep and shallow water, and at intermediate depths:

Deep-water Characteristics and Opportunities

Deep water corresponds with a water depth larger than half the wavelength, which is the common situation in the sea and ocean. In deep water, longer-period waves propagate faster and transport their energy faster. The deep-water group velocity is half the phase velocity. In shallow water, for wavelengths larger than about twenty times the water depth, as found quite often near the coast, the group velocity is equal to the phase velocity.

History

The first known patent to use energy from ocean waves dates back to 1799, and was filed in Paris by Girard and his son. An early application of wave power was a device constructed around 1910 by Bochaux-Praceique to light and power his house at Royan, near Bordeaux in France. It appears that this was the first oscillating water-column type of wave-energy device. From 1855 to 1973 there were already 340 patents filed in the UK alone.

Modern scientific pursuit of wave energy was pioneered by Yoshio Masuda's experiments in the 1940s. He has tested various concepts of wave-energy devices at sea, with several hundred units used to power navigation lights. Among these was the concept of extracting power from the angular motion at the joints of an articulated raft, which was proposed in the 1950s by Masuda.

A renewed interest in wave energy was motivated by the oil crisis in 1973. A number of university researchers re-examined the potential to generate energy from ocean waves, among whom notably were Stephen Salter from the University of Edinburgh, Kjell Budal and Johannes Falnes from Norwegian Institute of Technology (now merged into Norwegian University of Science and Technology), Michael E. McCormick from U.S. Naval Academy, David Evans from Bristol University, Michael French from University of Lancaster, Nick Newman and C. C. Mei from MIT.

Stephen Salter's 1974 invention became known as Salter's duck or *nodding duck*, although it was officially referred to as the Edinburgh Duck. In small scale controlled tests, the Duck's curved cam-like body can stop 90% of wave motion and can convert 90% of that to electricity giving 81% efficiency.

In the 1980s, as the oil price went down, wave-energy funding was drastically reduced. Nevertheless, a few first-generation prototypes were tested at sea. More recently, following the issue of climate change, there is again a growing interest worldwide for renewable energy, including wave energy.

The world's first marine energy test facility was established in 2003 to kick start the development of a wave and tidal energy industry in the UK. Based in Orkney, Scotland, the European Marine Energy Centre (EMEC) has supported the deployment of more wave and tidal energy devices than at any other single site in the world. EMEC provides a variety of test sites in real sea conditions. It's grid connected wave test site is situated at Billia Croo, on the western edge of the Orkney mainland, and is subject to the full force of the Atlantic Ocean with seas as high as 19 metres recorded at the site. Wave energy developers currently testing at the centre include Aquamarine Power, Pelamis Wave Power, ScottishPower Renewables and Wello.

Modern Technology

Wave power devices are generally categorized by the method used to capture the energy of the waves, by location and by the power take-off system. Locations are shoreline, nearshore and offshore. Types of power take-off include: hydraulic ram, elastomeric hose pump, pump-to-shore, hydroelectric turbine, air turbine, and linear electrical generator. When evaluating wave energy as a technology type, it is important to distinguish between the four most common approaches: point absorber buoys, surface attenuators, oscillating water columns, and overtopping devices.

Generic wave energy concepts: 1. Point absorber, 2. Attenuator, 3. Oscillating wave surge converter, 4. Oscillating water column, 5. Overtopping device, 6. Submerged pressure differential

Point Absorber Buoy

This device floats on the surface of the water, held in place by cables connected to the seabed. Buoys use the rise and fall of swells to drive hydraulic pumps and generate electricity. EMF generated by electrical transmission cables and acoustics of these devices may be a concern for marine organisms. The presence of the buoys may affect fish, marine mammals, and birds as potential minor collision risk and roosting sites. Potential also exists for entanglement in mooring lines. Energy removed from the waves may also affect the shoreline, resulting in a recommendation that sites remain a considerable distance from the shore.

Surface Attenuator

These devices act similarly to point absorber buoys, with multiple floating segments connected to one another and are oriented perpendicular to incoming waves. A flexing motion is created by swells that drive hydraulic pumps to generate electricity. Environmental effects are similar to those of point absorber buoys, with an additional concern that organisms could be pinched in the joints.

Oscillating Wave Surge Converter

These devices typically have one end fixed to a structure or the seabed while the other end is free to move. Energy is collected from the relative motion of the body compared to the fixed point. Oscillating wave surge converters often come in the form of floats, flaps, or membranes. Environmental concerns include minor risk of collision, artificial reefing near the fixed point, EMF effects from subsea cables, and energy removal effecting sediment transport. Some of these designs incorporate parabolic reflectors as a means of increasing the wave energy at the point of capture. These capture systems use the rise and fall motion of waves to capture energy. Once the wave energy is captured at a wave source, power must be carried to the point of use or to a connection to the electrical grid by transmission power cables.

Oscillating Water Column

Oscillating Water Column devices can be located on shore or in deeper waters offshore. With an air chamber integrated into the device, swells compress air in the chambers forcing air through an air turbine to create electricity. Significant noise is produced as air is pushed through the turbines, potentially affecting birds and other marine organisms within the vicinity of the device. There is also concern about marine organisms getting trapped or entangled within the air chambers.

Overtopping Device

Overtopping devices are long structures that use wave velocity to fill a reservoir to a greater water level than the surrounding ocean. The potential energy in the reservoir height is then captured with low-head turbines. Devices can be either on shore or floating offshore. Floating devices will have environmental concerns about the mooring system affecting benthic organisms, organisms becoming entangled, or EMF effects produced from subsea cables. There is also some concern regarding low levels of turbine noise and wave energy removal affecting the nearfield habitat.

List of Devices

A more complete list of wave energy developers is maintained here: Wave energy developers

Environmental Effects

Common environmental concerns associated with marine energy developments include:

- The risk of marine mammals and fish being struck by tidal turbine blades;

- The effects of EMF and underwater noise emitted from operating marine energy devices;

- The physical presence of marine energy projects and their potential to alter the behavior of marine mammals, fish, and seabirds with attraction or avoidance;

- The potential effect on nearfield and farfield marine environment and processes such as sediment transport and water quality.

The Tethys database provides access to scientific literature and general information on the potential environmental effects of wave energy.

Potential

The worldwide resource of wave energy has been estimated to be greater than 2 TW. Locations with the most potential for wave power include the western seaboard of Europe, the northern coast of the UK, and the Pacific coastlines of North and South America, Southern Africa, Australia, and New Zealand. The north and south temperate zones have the best sites for capturing wave power. The prevailing westerlies in these zones blow strongest in winter.

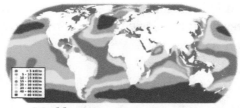

World wave energy resource map

Challenges

There is a potential impact on the marine environment. Noise pollution, for example, could have negative impact if not monitored, although the noise and visible impact of each design vary greatly. Other biophysical impacts (flora and fauna, sediment regimes and water column structure and flows) of scaling up the technology is being studied. In terms of socio-economic challenges, wave farms can result in the displacement of commercial and recreational fishermen from productive fishing grounds, can change the pattern of beach sand nourishment, and may represent hazards to safe navigation. Waves generate about 2,700 gigawatts of power. Of those 2,700 gigawatts, only about 500 gigawatts can be captured with the current technology.

Wave Farms

Portugal

- The Aguçadoura Wave Farm was the world's first wave farm. It was located 5 km (3 mi) offshore near Póvoa de Varzim, north of Porto, Portugal. The farm was designed to use three Pelamis wave energy converters to convert the motion of the ocean surface waves into electricity, totalling to 2.25 MW in total installed capacity. The farm first generated electricity in July 2008 and was officially opened on September 23, 2008, by the Portuguese Minister of Economy. The wave farm was shut down two months after the official opening in November 2008 as a result of the financial collapse of Babcock & Brown due to the global economic crisis. The machines were off-site at this time due to technical problems, and although resolved have not returned to site and were subsequently scrapped in 2011 as the technology had moved on to the P2 variant as supplied to Eon and Scottish Power Renewables. A second phase of the project planned to increase the installed capacity to 21 MW using a further 25 Pelamis machines is in doubt following Babcock's financial collapse.

United Kingdom

- Funding for a 3 MW wave farm in Scotland was announced on February 20, 2007, by the Scottish Executive, at a cost of over 4 million pounds, as part of a £13 million funding package for marine power in Scotland. The first machine was launched in May 2010.

- A facility known as Wave hub has been constructed off the north coast of Cornwall, England, to facilitate wave energy development. The Wave hub will act as giant extension cable, allowing arrays of wave energy generating devices to be connected to the electricity grid. The Wave hub will initially allow 20 MW of capacity to be connected, with potential expansion to 40 MW. Four device manufacturers have so far expressed interest in connecting to the Wave hub. The scientists have calculated that wave energy gathered at Wave Hub will be enough to power up to 7,500 households. The site has the potential to save greenhouse gas emissions of about 300,000 tons of carbon dioxide in the next 25 years.

Australia

- A CETO wave farm off the coast of Western Australia has been operating to prove commercial viability and, after preliminary environmental approval, underwent further development. In early 2015 a $100 million, multi megawatt system was connected to the grid, with all the electricity being bought to power HMAS Stirling naval base. Two fully submerged buoys which are anchored to the seabed, transmit the energy from the ocean swell through hydraulic pressure onshore; to drive a generator for electricity, and also to produce fresh water. As of 2015 a third buoy is planned for installation.

- Ocean Power Technologies (OPT Australasia Pty Ltd) is developing a wave farm connected to the grid near Portland, Victoria through a 19 MW wave power station. The project has received an AU $66.46 million grant from the Federal Government of Australia.

- Oceanlinx will deploy a commercial scale demonstrator off the coast of South Australia at Port MacDonnell before the end of 2013. This device, the *greenWAVE*, has a rated electrical capacity of 1MW. This project has been supported by ARENA through the Emerging Renewables Program. The *greenWAVE* device is a bottom standing gravity structure, that does not require anchoring or seabed preparation and with no moving parts below the surface of the water.

United States

- Reedsport, Oregon – a commercial wave park on the west coast of the United States located 2.5 miles offshore near Reedsport, Oregon. The first phase of this project is for ten PB150 Power-Buoys, or 1.5 megawatts. The Reedsport wave farm was scheduled for installation spring 2013. In 2013, the project has ground to a halt because of legal and technical problems.

- Kaneohe Bay Oahu, Hawai - Navy's Wave Energy Test Site (WETS) currently testing the Azura wave power device The Azura wave power device is 45-ton wave energy converter located at a depth of 30 meters in Kaneohe Bay.

Patents

- U.S. Patent 8,806,865 — 2011 *Ocean wave energy harnessing device* – Pelamis/Salter's Duck Hybrid patent

- U.S. Patent 3,928,967 — 1974 *Apparatus and method of extracting wave energy* – The original "Salter's Duck" patent

- U.S. Patent 4,134,023 — 1977 *Apparatus for use in the extraction of energy from waves on water* – Salter's method for improving "duck" efficiency

- U.S. Patent 6,194,815 — 1999 *Piezoelectric rotary electrical energy generator*

- Wave energy converters utilizing pressure differences US 20040217597 A1 — 2004 *Wave energy converters utilizing pressure differences*

References

- Takahashi, M.M.; Translated by: Kitazawa, K. and Snowden, P. (2000) [1991]. Deep Ocean Water as Our Next Natural Resource. Tokyo, Japan: Terra Scientific Publishing Company. ISBN 4-88704-125-X.

- Tucker, M.J.; Pitt, E.G. (2001). "2". In Bhattacharyya, R.; McCormick, M.E. Waves in ocean engineering (1st ed.). Oxford: Elsevier. pp. 35–36. ISBN 0080435661.

- Holthuijsen, Leo H. (2007). Waves in oceanic and coastal waters. Cambridge: Cambridge University Press. ISBN 0-521-86028-8.

- R. G. Dean & R. A. Dalrymple (1991). Water wave mechanics for engineers and scientists. Advanced Series on Ocean Engineering. 2. World Scientific, Singapore. ISBN 978-981-02-0420-4. See page 64–65.

- Cox, S. (10 July 2012). "Cooling a Warming Planet: A Global Air Conditioning Surge". Environment360. Retrieved 14 November 2016.

- Graham, Karen." First wave-produced power in U.S. goes online in Hawaii" Digital Journal. 19 September 2016. Web Accessed 22 September 2016.

- http://otecorporation.com/2016/08/30/ocean-thermal-energy-corporation-reports-announcement-bahamian-government-remobilization-completion-opening-baha-mar-beach-resort/

- Coxworth, B. (November 26, 2010). "More funds for Hawaii's Ocean Thermal Energy Conversion plant". Retrieved 14 November 2016.

- "Implementing Agreement on Ocean Energy Systems (IEA-OES), Annual Report 2007" (PDF). International Energy Agency, Jochen Bard ISET. 2007. p. 5. Archived from the original on 1 July 2015.

- Aalbers, R.R.D. (2015). Life cycle assessment of ocean thermal energy conversion - The life cycle impact of electricity supply on small island regions (MSc.). Delft University of Technology.

- Gunn, Kester; Stock-Williams, Clym (August 2012). "Quantifying the global wave power resource". Renewable Energy. Elsevier. 44: 296–304. doi:10.1016/j.renene.2012.01.101. Retrieved 27 February 2015.

- "WA wave energy project turned on to power naval base at Garden Island". ABC News Online. Australian Broadcasting Corporation. 18 February 2015. Retrieved 20 February 2015.

- Downing, Louise (February 19, 2015). "Carnegie Connects First Wave Power Machine to Grid in Australia". BloombergBusiness. Bloomberg. Retrieved 20 February 2015.

- Rajagopalan, K.; Nihous, G.C. (2013). "Estimates of Global Ocean Thermal Energy Conversion (OTEC) Resources Using an Ocean General Circulation Model". 50. Renewable Energy.

- Vega, L.A. and Comfort, C. "Environmental Assessment of Ocean Thermal Energy Conversion in Hawaii" (PDF). Hawaii National Marine Renewable Energy Center. Retrieved 27 March 2013.

- Rocheleau, G.; Grandelli, P. (22 September 2011). "Physical and biological modeling of a 100 megawatt Ocean Thermal Energy Conversion discharge plume". Institute of Electrical and Electronics Engineers: 3. Retrieved 27 March 2013.

- "Ocean Thermal Energy Conversion: Assessing Potential Physical, Chemical, and Biological Impacts and Risks" (PDF). National Oceanic and Atmospheric Administration, Office of Ocean and Coastal Resource Management. Retrieved 27 March 2013.

- Daly, J. (December 5, 2011). "Hawaii About to Crack Ocean Thermal Energy Conversion Roadblocks?". Oil-Price.com. Retrieved 28 March 2013.

Permissions

All chapters in this book are published with permission under the Creative Commons Attribution Share Alike License or equivalent. Every chapter published in this book has been scrutinized by our experts. Their significance has been extensively debated. The topics covered herein carry significant information for a comprehensive understanding. They may even be implemented as practical applications or may be referred to as a beginning point for further studies.

We would like to thank the editorial team for lending their expertise to make the book truly unique. They have played a crucial role in the development of this book. Without their invaluable contributions this book wouldn't have been possible. They have made vital efforts to compile up to date information on the varied aspects of this subject to make this book a valuable addition to the collection of many professionals and students.

This book was conceptualized with the vision of imparting up-to-date and integrated information in this field. To ensure the same, a matchless editorial board was set up. Every individual on the board went through rigorous rounds of assessment to prove their worth. After which they invested a large part of their time researching and compiling the most relevant data for our readers.

The editorial board has been involved in producing this book since its inception. They have spent rigorous hours researching and exploring the diverse topics which have resulted in the successful publishing of this book. They have passed on their knowledge of decades through this book. To expedite this challenging task, the publisher supported the team at every step. A small team of assistant editors was also appointed to further simplify the editing procedure and attain best results for the readers.

Apart from the editorial board, the designing team has also invested a significant amount of their time in understanding the subject and creating the most relevant covers. They scrutinized every image to scout for the most suitable representation of the subject and create an appropriate cover for the book.

The publishing team has been an ardent support to the editorial, designing and production team. Their endless efforts to recruit the best for this project, has resulted in the accomplishment of this book. They are a veteran in the field of academics and their pool of knowledge is as vast as their experience in printing. Their expertise and guidance has proved useful at every step. Their uncompromising quality standards have made this book an exceptional effort. Their encouragement from time to time has been an inspiration for everyone.

The publisher and the editorial board hope that this book will prove to be a valuable piece of knowledge for students, practitioners and scholars across the globe.

Index

A

Air Conditioning, 152, 163, 292, 301
Arch Dams, 9-10, 12-14, 29-30, 32
Arch Style Diversion Dam, 43
Arch-gravity Dams, 9, 12, 15

B

Barrages, 15, 17, 174-175, 178, 191, 193-195
Base Load Power Plant, 123, 146
Beaver Dams, 21
Buttress Style Diversion Dam, 43

C

Cantilever Strutted, 46-47
Check Dam, 18, 33-37
Cofferdam, 20-21, 28, 37-39, 46, 81, 97
Cofferdams, 20-21, 38-39, 86, 97
Compressed Air Hydro, 3
Concrete-face Rock-fill Dams, 16
Conduit Hydroelectricity, 3, 134
Control Systems, 132, 198, 251
Conventional Hydroelectric, 3, 112, 116, 123, 153, 167, 173

D

Dam, 1, 3-114, 116-127, 129, 131-132, 136, 142, 145-148, 150, 152, 162, 168, 170, 174-175, 179-181, 191-192, 195, 204, 208-209, 220, 237, 246, 248, 253, 255-256, 258-260, 267
Dam Creation, 23
Dam Failure, 26-27, 119
Decentralised Systems, 170
Detention Dam, 39-41
Direct Pumping, 169
Direct Strutted, 46
Distributed Generation, 126, 161
Diversion Dam, 8, 15, 20, 23, 41-43
Diversionary Dam, 18
Drainage Ditch, 33
Dry Dam, 18

E

Earth-fill Dams, 16-17
Earth-filled Dam, 17, 44, 74
Ecosystem Damage, 117
Embankment Dams, 9, 12, 15, 19, 28, 42, 44

Embankment Style Diversion Dam, 42
Energy Storage, 110, 113, 124, 127, 129, 145, 149-155, 157, 159-167, 170-171, 175, 259

F

Flood Detention Dam, 39-41
Flow Shortage, 118

G

Gorlov, 132, 185, 250, 266-267, 269-270
Grade Control Mechanism, 34
Gravitational Potential Energy, 150-151, 154, 166
Gravity Dam, 8, 12, 14-15, 27-28, 62, 91, 93
Gravity Dams, 9, 11-12, 14-15, 28, 43
Gravity Style Diversion Dam, 43
Grid Electricity, 162

H

Home Energy Storage, 161
Hydraulic Mining, 2, 247
Hydraulic Power-pipe Networks, 2
Hydroelectric Power, 1, 3, 12, 18, 21-23, 26, 39, 42, 53, 74-75, 80-81, 91, 111-115, 117-118, 121, 125-127, 130, 134, 136-137, 142, 146-149, 240, 270, 278
Hydroelectricity, 1, 3, 22, 49, 51, 53, 55, 57, 59, 61, 63, 65, 67, 69, 71, 73, 75, 77, 79, 81, 83, 85, 87, 89, 91, 93, 95, 97, 99, 101, 103, 105, 107, 109-111, 113-123, 125, 127, 129, 131, 133-135, 137, 139, 141-143, 145, 147-149, 151-153, 155, 157, 159, 161-163, 165-167, 169, 171, 249, 276
Hydroplants, 135
Hydropower Sustainability, 5-6
Hydrostatic Pressure, 10, 12, 29

I

Impulse Turbines, 245, 248, 250, 271

K

Kaplan Turbine, 132, 211, 244, 247, 249, 251-252, 259-262

L

Load Balancing, 166

M

Mass Concrete Dams, 28
Mechanical Storage, 152
Methane Emissions, 118

Micro Hydro, 3, 110, 115, 120, 127, 130, 132-133, 267

N
Natural Dams, 21

P
Peaking Power Plant, 123, 147 Pico Hydro, 110, 115, 127, 130, 132, 134-135

Portable Cofferdams, 39

Potential Energy, 5, 22, 112, 124, 150-151, 154-155, 161, 166, 169, 174-175, 195-196, 238, 240, 242, 244, 248, 256, 272, 274, 284-285, 287, 295, 298

Power Generation, 3, 21-24, 26, 39, 42, 52, 65, 74-75, 77-78, 81, 93, 103-104, 115-116, 120, 125, 134-135, 139, 179, 181-183, 189, 191, 195-196, 198, 207, 209, 240, 245, 249, 278-281, 283, 293

Power Generation Plant, 21

Pressure Reducing Valves, 134

Propeller Turbines, 132, 262

Pumped Hydroelectric Energy Storage (phes), 166

Pumped-storage, 3, 22, 112-113, 127, 142, 145, 148-149, 151, 153, 162-163, 166-170, 249

Pumped-storage Hydroelectricity, 3, 22, 145, 151, 153, 162, 166-167, 249

Pumped-storage Hydropower, 127, 170

R
Reaction Turbines, 248-250

Rechargeable Battery, 150-151, 156, 161, 165

Relocation, 25-26, 61, 67, 86, 88, 119

Renewable Electricity, 110-111, 120

Renewable Energy, 1-2, 21, 114, 120, 127, 142, 144, 151, 162-163, 170, 172-173, 175, 182, 185, 188, 191, 196, 200, 208, 253, 276, 279, 282, 286-287, 296, 301

Rock-fill Dams, 16

Rock-filled Dam, 44

Run-of-the-river Hydroelectricity, 3, 110, 122-123

S
Saddle Dam, 17

Scalloping, 47

Screw Turbine, 132, 250-251

Sea Water, 169, 195, 242, 282, 284, 286, 292

Sediment Detention Dam, 40-41

Seepage Erosion, 44

Siltation, 118

Small Hydro, 3-4, 110, 114, 126-129, 246, 274

Spillways, 22-23, 45, 47, 62, 72-75, 77, 104-106, 114

Steel Dam, 19, 45-47

Steel Dams, 19, 46-47

T
Tailings Dam, 19

Thermal Storage, 155, 161, 163-164

Timber Dams, 20

Transmission Distance, 135

Turbine Blade Materials, 252

U
Underground Dam, 18

Underground Power Station, 110, 115, 147-149

Underground Reservoirs, 170

W
Water Power, 1-2, 171, 214, 226, 229-230, 255

Water Quality Control Mechanism, 35

Water Turbines, 49, 132, 134-135, 142, 150, 168, 217, 227, 231, 233, 243-249, 252-254, 263

Water Wheel, 2, 9, 133, 211, 218-220, 223-239, 241, 243, 245-246, 250-251, 254-255, 263-264, 271, 275

Watermills, 1, 211-212, 215-219, 221-222, 227-228, 233-234, 240

World Hydroelectric Capacity, 120